GW00759820

SAFETY RISK MANAGEMENT FOR MEDICAL DEVICES

SAFETY RISK MANAGEMENT FOR MEDICAL DEVICES

BIJAN ELAHI
Technical Fellow, Medtronic, The Netherlands

ELSEVIER

ACADEMIC PRESS
An imprint of Elsevier

Academic Press is an imprint of Elsevier
125 London Wall, London EC2Y 5AS, United Kingdom
525 B Street, Suite 1650, San Diego, CA 92101, United States
50 Hampshire Street, 5th Floor, Cambridge, MA 02139, United States
The Boulevard, Langford Lane, Kidlington, Oxford OX5 1GB, United Kingdom

Copyright © 2018 Elsevier Ltd. All rights reserved.

No part of this publication may be reproduced or transmitted in any form or by any means, electronic or mechanical, including photocopying, recording, or any information storage and retrieval system, without permission in writing from the publisher. Details on how to seek permission, further information about the Publisher's permissions policies and our arrangements with organizations such as the Copyright Clearance Center and the Copyright Licensing Agency, can be found at our website: www.elsevier.com/permissions.

This book and the individual contributions contained in it are protected under copyright by the Publisher (other than as may be noted herein).

Notices
Knowledge and best practice in this field are constantly changing. As new research and experience broaden our understanding, changes in research methods, professional practices, or medical treatment may become necessary.

Practitioners and researchers must always rely on their own experience and knowledge in evaluating and using any information, methods, compounds, or experiments described herein. In using such information or methods they should be mindful of their own safety and the safety of others, including parties for whom they have a professional responsibility.

To the fullest extent of the law, neither the Publisher nor the authors, contributors, or editors, assume any liability for any injury and/or damage to persons or property as a matter of products liability, negligence or otherwise, or from any use or operation of any methods, products, instructions, or ideas contained in the material herein.

British Library Cataloguing-in-Publication Data
A catalogue record for this book is available from the British Library

Library of Congress Cataloging-in-Publication Data
A catalog record for this book is available from the Library of Congress

ISBN 978-0-12-813098-8

For Information on all Academic Press publications
visit our website at https://www.elsevier.com/books-and-journals

ELSEVIER
Book Aid International
Working together to grow libraries in developing countries
www.elsevier.com • www.bookaid.org

Publisher: Mara Conner
Acquisition Editor: Fiona Geraghty
Editorial Project Manager: Katie Chan
Production Project Manager: Vijay R. Bharath
Cover Designer: Matthew Limbert

Typeset by MPS Limited, Chennai, India

DEDICATION

This book is dedicated to all the people of this planet who suffer from illness and who look to medical technology with hope and trust.

CONTENTS

LIST OF FIGURES

LIST OF TABLES

BIOGRAPHY

Award winning, international educator and consultant

Bijan Elahi has worked in risk management for medical devices for over 25 years at the largest medical device companies in the world, as well as small startups. He is currently employed by Medtronic as a Technical Fellow where he serves as the corporate expert on product safety risk management. In this capacity, he offers consulting and education on risk management to all Medtronic business units, worldwide. He is also a lecturer at Eindhoven University of Technology in the Netherlands, where he teaches risk management to doctoral students in engineering. He is a frequently invited speaker at professional conferences, and is also a contributor to ISO 14971, the international standard on the application of risk management to medical devices.

PREFACE

A writer is, after all, only half his book. The other half is the reader and from the reader the writer learns.

P.L. Travers, author (1899–1996)

This book is written to serve both the professionals in the MedTech industry as well as the students in universities. The book delivers not only the theory but also offers practical guidance on how to apply the theory in your day-to-day work. The objective of this book is to demystify risk management and provide clarity of thought and confidence to the practitioners of risk management as they do their work.

I offer here the result of 30 + years of experience in risk management beginning with my work on the Space Shuttle at NASA and continuing in the medical device industry. What is presented in this book is the best available knowledge today. But as in any scientific or technical endeavor, the methods and techniques in risk management will continue to evolve, mature, and improve.

Although I serve the Medtronic corporation, the opinions and materials presented in this book are mine and do not represent the Medtronic corporation.

Bijan Elahi

CHAPTER 1

Introduction

Abstract

This book is about management of safety risks for medical devices. You will learn how to answer difficult questions such as: Is my medical device safe enough? What are the safety-critical aspects of my device, and which are the most important ones? Have I reduced the risks as far as possible? A sound and properly executed risk management process does not render a risk-free medical device. It does imply that the best efforts were made to produce a device that is adequately safe—a device that provides benefits which outweigh its risks.

Keywords: Risk management; medical device

This book is about management of safety risks for medical devices. You will learn how to answer difficult questions such as: Is my medical device safe enough? What are the safety-critical aspects of my device, and which are the most important ones? Have I reduced the risks as far as possible?

One of the main challenges of medical device companies is to be able to tell a clear and understandable story in their medical device submissions as to why their medical device is safe enough for commercial use. The methodology offered in this book, which we call the "BXM method" offers a simple, understandable, and integrated process that is scalable and efficient. The BXM method is suitable for automation, which is a huge benefit to the practitioners of risk management in an environment of dynamic and complex medical technology.

The methodology offered in this book is in conformance with ISO 14971, and has been used and tested numerous times in real products that have been submitted and approved by the FDA and European notified bodies.

Risk management is a truly interdisciplinary endeavor. A successful risk manager employs skills from engineering, physics, chemistry, mathematics, logic, behavioral science, psychology, and communication, to name a few. In this book, we touch upon ways in which the various disciplines are employed at the service of risk management.

This book is designed to serve both university students who are neophytes to risk management, as well as industry professionals who need a reference handbook. An example of execution of the BXM method on a hypothetical medical device is provided in Appendix C as a model to further elucidate the BXM method.

Safety Risk Management for Medical Devices
DOI: https://doi.org/10.1016/B978-0-12-813098-8.00001-5
© 2018 Elsevier Ltd.
All rights reserved.

Although the techniques, information, and tips that are offered in this book are intended for medical technology and devices, other safety-critical fields can also benefit from the knowledge gained herein.

One of the factors that should be kept in mind in the analysis of risks is that the risk is not a deterministic outcome, but rather a probabilistic phenomenon. The same therapy from a device could have different consequences for different patients. Variations in patient physiology and environmental conditions can contribute to vastly different severities of harm, from patient to patient.

Risk management before a product is launched is about predictive engineering—to forecast risks, and attempt to reduce and control the risks to acceptable levels. This is in contrast to postmarket risk management which takes the reactionary approach of root cause analysis, and Corrective and Preventive Actions after an adverse event has happened in the field.

Manufacturers are not expected to be error-free, flawless, or perfect. They *are* expected to use sound processes and good judgment to reduce the probability of harm to people.

A sound and properly executed risk management process does not render a risk-free medical device. It *does* imply that the best efforts were made to produce a device that is adequately safe—a device that provides benefits which outweigh its risks. Human beings are prone to errors and poor judgment. This is called misfeasance in legal terms. It is different from malfeasance, which is deliberate or deceptive actions intended to release a device that is not adequately safe.

Safety risk management is applicable to the entire lifecycle of medical devices including design, production, distribution, installation, use, service, maintenance, obsolescence, and decommissioning, and even destruction or disposal.

Although Harm is defined as "injury or damage to the health of people, or damage to property or the environment" [1], in this book, we focus on injury or damage to the health of people. Damage to property or the environment that has a direct impact to the health of people is also within the scope of this book.

The words: *System*, *Product*, and *medical device* are used interchangeably in this book to refer to the target of the risk management process.

Styling—In this book, words that have specific meaning in the world of risk management are capitalized to distinguish them from the ordinary dictionary meanings. Examples: Cause, End Effect, and Risk Control.

CHAPTER 2

Why Do Risk Management?

Abstract

There are many good reasons to do risk management. In addition to making safer products, risk management can help reduce the cost of design and development by identifying the safety-critical aspects of the design early in the product lifecycle. Risk management is a legal requirement in most countries, without which it would not be possible to obtain approval for commercialization of medical devices. In the unfortunate situations when people are injured by medical devices, the first place that lawyers would look is the risk management file of the device.

Keywords: Legal; Regulatory; risk-based; recall; field corrective action; moral; ethical

Whether you are aware of it or not, you are constantly managing risk in your daily life. For almost every action that we take, we internally evaluate the benefit of that action versus the risks (or cost) of that action. If we believe the benefits outweigh the risks, we take that action. Else, we don't. Consider the simple action of driving to work in your car. You consider the benefit of comfort and speed of getting from home to work, versus the risks of getting injured or killed in a car crash. In general, the chances of getting into a serious accident are fairly small, compared to the benefit of commuting in your car. But now imagine you are in a war-torn country where there are explosive devices buried in the roadway. Now the calculus changes. The risks are higher than the benefits, and you would likely choose to walk off-road instead.

The medical technology (MedTech) industry is required to evaluate the potential risks due to the use of a medical device against the potential benefits of that device. Regulatory approval of a medical device requires demonstrating that the risks of the device are outweighed by its benefits. Rather than making a subjective judgment, formal and systematic methods are used to make this determination.

Another very important reason to do risk management is the progressive shift in the industry where more and more decisions are risk based. Risk-based decisions are rational and defensible. In many aspects of product development, e.g., design choices, or sample size determination, risk is a good discriminator and basis for decision making. Moreover, the European Medical Device Regulation (EU MDR)

Safety Risk Management for Medical Devices
DOI: https://doi.org/10.1016/B978-0-12-813098-8.00002-7

© 2018 Elsevier Ltd.
All rights reserved.

[2] takes a preference for a risk-based approach to evaluation of manufacturer's technical documentation, and oversight and monitoring of the manufacturers. How can one make risk-based decisions, if one doesn't know the risks? Risk management offers the answer.

2.1 LEGAL AND REGULATORY REQUIREMENTS

2.1.1 United States

In the United States, the governing law is U.S. CFR Title 21, part 820. Title 21 is about foods and drugs, and part 820 is about Quality System Regulations. This law requires that all finished medical devices be safe and effective. The burden of proof is on the manufacturer. Prior to ISO 14971, there were many methods used by manufacturers to provide evidence of safety. There was no consistency and the quality of the evidence varied widely.

On June 27, 2016 the FDA recognized ISO 14971:2007 [3] as a suitable standard for risk management. Therefore compliance to ISO 14971:2007 [3] is sufficient proof of safety for the FDA.

2.1.2 European Union

The European Union Directive 93/42/EEC, also known as the Medical Device Directive (MDD) [4] compels the member States to pass laws that are consistent with the MDD. Article 3 of the MDD requires the medical devices must meet the essential requirements set out in Annex I. Stated briefly and simply, the Essential Requirements of Annex I stipulate that medical devices:

1. Be safe when used as intended by the manufacturer
2. Their risks be outweighed by their benefits
3. The risks be reduced as far as possible

Article 5 of the MDD states that compliance with the Essential Requirements of Annex I can be presumed, if a medical device is conformant with harmonized standards that are published in the *Official Journal of the European Communities* [5].

There is also a counterpart to MDD [4] for active implantable medical devices. It is called Active Implantable Medical Device Directive (AIMDD) [6].

One of the standards which is published in the *Official Journal of the European* [5] *Communities* is EN ISO 14971:2012 [7]. Therefore one can conclude that conformance to EN ISO 14971:2012 [7] is grounds for claiming compliance with the Essential Requirements of the MDD.

Each country in the European Union has a competent authority who approves medical devices for commercialization. Upon approval by a competent authority, a medical device can be CE (Conformité Européenne) marked:

CE

Notified bodies review submissions by the manufacturers, for compliance. Notified Bodies are accredited entities who assess conformity to harmonized standards. For a list of the Notified Bodies refer to the website: ec.europa.eu.

The new European Medical Device Regulation (EUMDR) [2] will eventually replace the MDD and AIMDD. The transition period started in May 2017.

2.1.3 MDD/AIMDD and transition to EU MDR

EU MDR [2] was promulgated on May 26, 2017. There is a 3-year transition period after which AIMDD [6] and MDD [4] will no longer be effective and only MDR certification will be possible (May 26, 2020). From May 26, 2017 to Nov/Dec 2018 only MDD/AIMDD certification is possible. Thereafter until May 26, 2020, it's possible to choose MDD/AIMDD or MDR [2] certification.

There is a 4-year grace period after May 26, 2020 during which products that were certified to MDD/AIMDD can be still manufactured and sold—until May 26, 2024. Thereafter, there is only a 1-year period until May 26, 2025 to sell off any inventory of MDD/AIMDD certified products.

2.2 BUSINESS REASONS

2.2.1 Cost efficiency

One of the main benefits of risk management is gaining knowledge of what the risks of a medical device are; where they are; and how big they are. With this knowledge, the product development group can focus their engineering resources on the areas of highest risk. Furthermore, good risk management practices can help detect design flaws which have a safety impact, early in the product development process. The sooner a design flaw is corrected, the less expensive it is to fix it.

Another factor that medical device manufacturers face is the concept of ALAP/AFAP, or reduction of risks to as low/far as possible. Without knowledge of the risks of the various aspects of the design, manufacturers may go overboard and overengineer the product particularly in the low/no risk areas of the design.

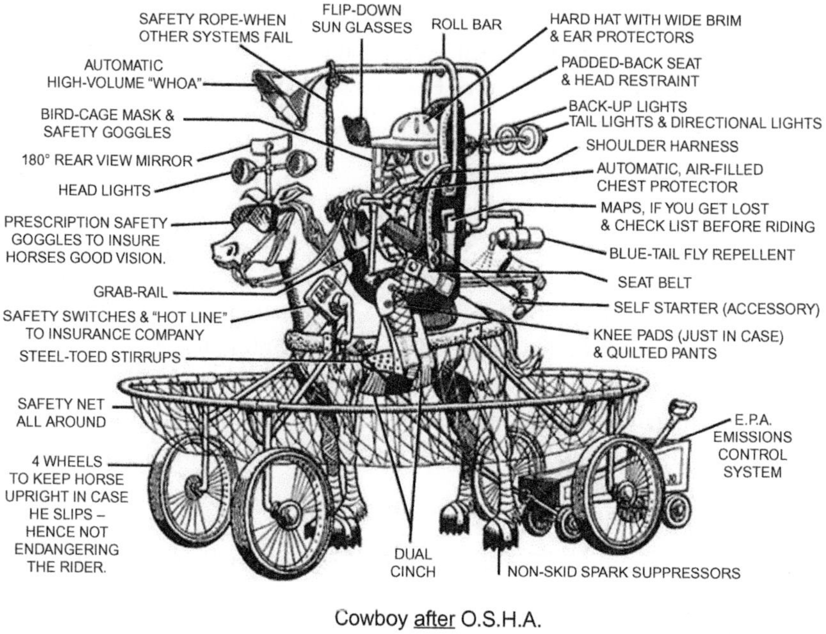

Figure 2.1 Overengineered cowboy.

The cartoon in Fig. 2.1 was drawn by J.N. Devin in 1972. Although it gives a comical view of overengineering, there are some lessons hidden in there. It shows when a good intention can go awry to the point of making a product useless.

2.2.2 Avoiding recalls and field corrective actions

Safety violations are the main reason for product recalls, and Field Corrective Actions. Product recalls are very expensive, and expose manufacturers to lawsuits and potentially large fines, settlement costs, and legal fees. Moreover, the reputation of a manufacturer may become tarnished and future sales hampered.

Good risk management practices can reduce the probability of harming people or the environment, and thus avoiding recalls.

One of the most important benefits of risk management is that it provides leading indicators for potential future problems. In many cases a manufacturer realizes only after an adverse event, that they are in trouble and facing a lawsuit, or punishment by Regulatory bodies. Risk management enables manufacturers to identify the highest risks associated with their products and be able to forecast the probability of serious adverse events.

2.2.3 Better communications

An unexpected side benefit of risk management is improved communication. In most companies the product-development teams become siloed, which means poor communication among the various disciplines, such as electrical engineering, mechanical engineering, clinical, sterilization, etc. Because risk management is a team effort, it tends to bring the various disciplines to the table to work together toward safer products. Many very useful and enlightening discussions happen during the risk management work meetings.

2.3 MORAL AND ETHICAL REASONS

Our patients trust us with their lives. They expect that we do our utmost to make devices that are as safe as possible. It is our moral and ethical duty to apply good risk management practices, so we deliver the safest possible products to our patients.

CHAPTER 3

The Basics

Abstract

As in any other discipline, risk management also has its own jargon, or special vocabulary. It is imperative that you learn this vocabulary and use it correctly and consistently. Without this common language, it is not possible to reliably convey meaning. This increases the probability of miscommunication, confusion and the potential for errors.

Keywords: Vocabulary; reasonably foreseeable misuse; hazard theory; integral system; distributed system

3.1 VOCABULARY OF RISK MANAGEMENT

As in any other discipline, risk management also has its own jargon, or special vocabulary. It is imperative that you learn this vocabulary and use it correctly and consistently. Just as important, you should teach this vocabulary to others who participate in producing risk management work products. Without this common language, it is not possible to reliably convey meaning. This increases the probability of miscommunication, confusion and the potential for errors. Table 3.1 lists some of the most commonly used terms in medical device risk management.

Table 3.1 Special vocabulary of risk management

Term	Definition
Basic Safety	freedom from unacceptable risk directly caused by physical hazards when ME equipment is used under normal condition and single fault condition [8]
Essential Performance	performance of a clinical function, other than that related to basic safety, where loss or degradation beyond the limits specified by the manufacturer results in an unacceptable risk [8]
Expected Service Life	time period specified by the manufacturer during which the medical electrical equipment or medical electrical system is expected to remain safe for use (i.e. maintain basic safety and essential performance) NOTE: Maintenance can be necessary during the expected service life. [8]
Failure	the inability of an entity to achieve its purpose

(Continued)

DOI: https://doi.org/10.1016/B978-0-12-813098-8.00003-9

© 2018 Elsevier Ltd.
All rights reserved.

Table 3.1 (Continued)

Term	Definition
Fault	an anomalous condition for a part
Harm	injury or damage to the health of people, or damage to property or the environment [1]
Hazard	potential source of harm [1]
Hazardous Situation	circumstance in which people, property, or the environment are exposed to one or more hazards [1]
Instruction for Use	information provided by the manufacturer to inform the user of the device's intended purpose and proper use and of any precautions to be taken [2] article 2, (14)
Intended Use	use for which a product, process or service is intended according to the specifications, instructions and information provided by the manufacturer [3]
Intended purpose	use for which a device is intended according to the data supplied by the manufacturer on the label, in the instructions for use or in promotional or sales materials or statements and as specified by the manufacturer in the clinical evaluation [2] Article 2 (12)
Label	written, printed or graphic information appearing either on the device itself, or on the packaging of each unit or on the packaging of multiple devices [2] Article 2 (13)
Residual Risk	risk remaining after risk reduction measures have been implemented [1] including actions to avoid, or limit the harm
Risk	combination of the probability of occurrence of harm and the severity of that harm [2] article 2, (23)
Risk Analysis	systematic use of available information to identify hazards and to estimate the risk [1]
Risk Assessment	overall process comprising a risk analysis and a risk evaluation [1]
Risk Control	process in which decisions are made and measures implemented by which risks are reduced to, or maintained within, specified levels [3]
Risk Estimation	process used to assign values to the probability of occurrence of harm and the severity of that harm [3]
Risk Evaluation	process of the estimated risk against given risk criteria to determine the acceptability of the risk [3]
Risk Management	systematic application of management policies, procedures and practices to the tasks of analyzing, evaluating, controlling and monitoring risk [3]
Risk Management File	set of records and other documents that are produced by risk management [3]
Safety	freedom from unacceptable risk [3]

(*Continued*)

Table 3.1 (Continued)

Term	Definition
Serious Injury	injury or illness that: [9] a. is life threatening, b. results in permanent impairment of a body function or permanent damage to a body structure, or c. necessitates medical or surgical intervention to prevent permanent impairment of a body function or permanent damage to a body structure NOTE Permanent impairment means an irreversible impairment or damage to a body structure or function excluding trivial impairment or damage.
System	a combination of products, either packaged together or not, which are intended to be inter-connected or combined to achieve a specific medical purpose [2] article 2 (11)
User	any healthcare professional or lay person who uses a device [2] article 2, (37)

3.1.1 Further elaborations

Harm – Although not explicitly stated in the Standard [3,7] the intention behind including damage to property and the environment in the scope of Harm, is to consider the type of damage that could have safety consequences. For example, improper disposal of radioactive isotopes in a brachytherapy device may endanger sanitation workers. In addition, with today's environment of cybersecurity concerns, data should be included in the scope of 'property'.

Safety – [1] advises to avoid the use of the terms "safety" and "safe" as descriptive adjectives because they could be misinterpreted as an assurance of freedom from risk. For example, "safety helmet" may mislead a person that wearing a safety helmet will protect the wearer from all head injuries. All medical devices carry a certain amount of residual risk, and the users should be made aware of such residual risks.

Risk – Although the definition of risk is simply "combination of the probability of occurrence of harm and the severity of that harm", there are many factors that play a role in the level of risk which is experienced by people. For example, exposure to a hot object causes burns. But it matters how hot the object is, how long the hot object contacts the person, where on/in the body the hot object contacts the person, and the physical properties of the hot object – compare a hot spoon vs. hot oil. Also, typically when a Harm happens, actions are taken to ameliorate the Harm. ISO/IEC Guide 51 [1], 3.9, Note 1, advises that in risk calculation the possibility to avoid or limit the Harm should be included.

Hazard Analysis vs. Risk Analysis – Sometimes, the terms 'Hazard Analysis' and 'Risk Analysis' are used interchangeably. This is incorrect. The purpose of hazard analysis is the identification of Hazards and the foreseeable sequence of events that could realize those Hazards. In contrast, risk analysis is about estimation of the potential risks due to the identified Hazards. Hazard analysis precedes risk analysis and identifies the Hazards. Risk analysis estimates the risks of Harms that could ensue from the identified Hazards.

3.1.1.1 Reasonably foreseeable misuse

It is a requirement of ISO 14971 [3,7] for the manufacturer to identify and document the intended use, and reasonably foreseeable misuses of the medical device. But the concept of reasonably foreseeable misuse creates confusion and debate due to its ambiguity. First, what is reasonable? In whose judgement? Second, what does foreseeable mean? Should every wild and creative idea be considered as reasonably foreseeable?

[1] offers this definition for reasonably foreseeable misuse: "use of a product or system in a way not intended by the supplier, but which can result from readily predictable human behavior". But this definition still leaves the ambiguity of the word 'readily'.

Misuse is defined as incorrect or improper use of the medical device by the Standard [3,7]. But how do you distinguish use-failure from misuse?

Below, some guidance is offered to help the analyst distinguish what is a reasonably foreseeable misuse. First, we distinguish misuse from use failure. Use failure is a failure of the user to perform a user-action on the medical device. This type of event is covered in the UMFMEA and its risks are accounted for, in the RACT. The following six tests are offered as a means to determine if something is a reasonably foreseeable misuse.

1. Deliberate
   There is a deliberate decision by the user to use the device in the manner they want.
2. Well-intentioned
   The User intends to do well by the patient, i.e., no harm is intended.
3. Beneficial
   The User believes a benefit can be derived for the patient from the misuse.
4. Feasible
   The misuse is feasible, i.e., it is technically, financially, and skill-wise within the capability of the user to do the misuse.
5. Safe
   The user can safely use the device in the manner that they wish to use the device.
6. Ethical
   The user is acting ethically. They have disclosed the truth about the misuse and have the consent of the patient and the hospital. (This is not relevant when a patient is using the device on him/herself.)

If a foreseen misuse meets the above six tests, then it can be construed as a reasonably foreseeable misuse. Off-label use is a typical kind of misuse. Malice is excluded from the analysis. That is, if the user intends to harm a patient, such action is not included in the risk management of the medical device.

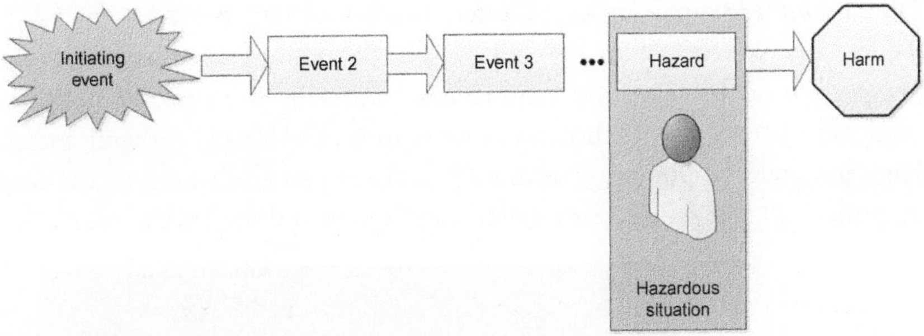

Figure 3.1 Hazard Theory.

It is a very good idea to consult with other departments such as sales, marketing and clinical staff to get insights into how the device might get misused in the field.

3.2 HAZARD THEORY

In order to receive Harm, there must be exposure to Hazard(s). Fig. 3.1 illustrates a model that is called Hazard Theory. The model states that an initiating event starts a progression of events that culminate in a Hazard. The Hazard is the last stop in the chain of events that lead to the Hazard. Unless there is exposure to the Hazard(s), there can be no Harm. Of course, Harm can be of various degrees of severity; anywhere from negligible to catastrophic.

In the BXM method the complete spectrum of Harm is considered. That is, given a Hazardous Situation, everything from nothing to death is considered. Therefore, it can be said that once the Hazardous Situation has been achieved, the probability of receiving Harm is 100%.

3.3 SYSTEM TYPES

The Systems that are subjected to the risk management process can be classified into two categories:

a. Integral Systems
These are Systems that are observable as <u>one</u> integral piece from the perspective of the user. They do not require any assembly or integration by the user. Example: a blood glucose monitor.

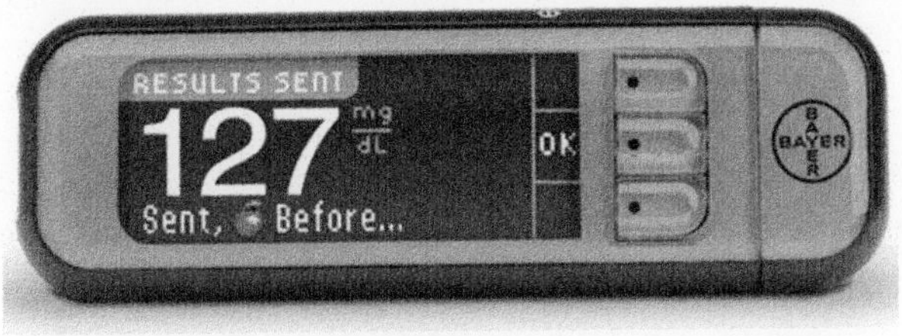

b. Distributed Systems

These are Systems that are comprised of multiple components from the perspective of the user, and require integration by the user. Example: a spinal cord stimulation System shown below with 6 individually integral components. Each integral component is separately packaged and delivered to the user. Final assembly and integration into a working System is done by the user.

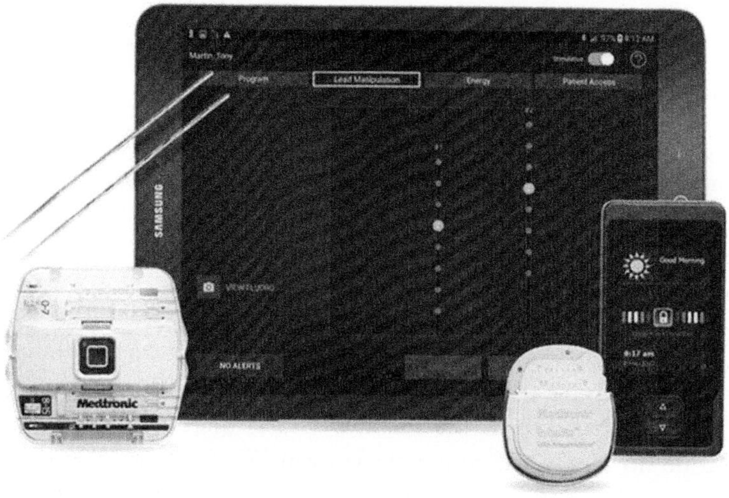

CHAPTER 4

Understanding Risk

Abstract

The word "risk" derives from the early Italian *risicare*, which means "to dare." The modern concept of risk and the associated mathematics of probability and statistics were driven first by interest in gambling. In the 17th century Blaise Pascal, a French mathematician was one of the first to address the puzzle of probability of the outcomes of a game of chance. There are many types of risk. This book is focused on safety risks, and contributors to risk. A major contributor to risk is the human, be it the designer, builder, maintainer, or the user of the System. Assuming good intention, the main cause of human influence on risk is inconsistency between reality, and the mental model of reality that the human holds.

Keywords: Risk; contributors to risk; risk perception; mental models

The word "risk" derives from the early Italian *risicare*, which means "to dare" [10]. The modern concept of risk and the associated mathematics of probability and statistics were driven first by interest in gambling. In the 17th century Blaise Pascal, a French mathematician was one of the first to address the puzzle of probability of the outcomes of a game of chance. Pascal turned to Pierre de Fermat and together they created the theory of probability [10]. This meant that for the first-time people could make decisions not based on superstition, but based on numbers. Another interesting consequence of this discovery was the invention of insurance, which itself promoted risk-taking in business and advanced international commerce.

Risk answers the question: if A happens, what is the probability of B happening. We employ statistics as a tool to leverage our knowledge of the past to predict the future.

In many cases risk implies choice. That is, with a prediction of the probability of B, we choose whether to take action A. For example, making an investment while estimating the probability of a financial gain is a type of risk taking.

Ultimately, risk is about balancing benefit versus cost. Cost in the context of medical device risk management is injury or damage to health.

It's important to understand that risk cannot be eliminated—risk is managed. We cannot build a "safe" device; i.e., a device with zero risk. But we can manage the risk. Management of risk involves creating boundaries, and then engineering products such that they do not pose unacceptable levels of risk within those boundaries. Boundaries include: intended use, intended user, intended use environment, and indicated therapy. According to Ref. [1], in the absence of a specific declaration of

Safety Risk Management for Medical Devices
DOI: https://doi.org/10.1016/B978-0-12-813098-8.00004-0
© 2018 Elsevier Ltd.
All rights reserved.

intended use by the manufacturer, generally understood patterns of usage can construe the intended use.

Safety is an emergent system property. That is because users interact with the system, not with just parts of the system. For example, a car is a system. Users interact with a car-system for the benefit of transportation. A tire is a component of the car-system. A tire disintegration does not pose a safety risk, if it is sitting in storage. But a tire disintegration does pose a safety risk if it is part of the car-system.

4.1 RISK DEFINITIONS

There are a number of definitions for "risk." Let's examine some:

1. Combination of the probability of occurrence of Harm & the severity of that Harm [3,7]
2. The objectified uncertainty regarding the occurrence of an undesirable event [11]
3. The probability of occurrence of something undesirable
4. Probability of sustaining Harm in a Hazardous Situation

To provide beneficial information, risk needs to be measurable—quantitatively or qualitatively. Simply knowing there is a risk of Harm is not very useful. One needs to know the magnitude of risk to be able to make a sound decision. Consider disclosure of risk by pharmaceuticals. Pfizer describes the following risks for taking Lipitor, a common statin drug for fighting cholesterol:

- Diarrhea
- Upset stomach
- Muscle and joint pain
- Feel tired or weak
- Loss of appetite
- Upper belly pain
- Dark, amber-colored urine
- Yellowing of your skin or the whites of your eyes, etc.

There is no citation of the *probability of occurrence* for any of the above harms. Without that information, it is not possible to make a sound decision as to whether to take Lipitor or not. Consider if the probability of occurrence of a Harm were 99% versus 0.01%; wouldn't your decision be affected by that knowledge?

If quantitative data is not available, relative, qualitative information would also be helpful. For example, if Crestor, which is another popular statin drug claimed that relative to Lipitor their risks of side-effects are lower, a patient might choose Crestor over Lipitor.

4.2 TYPES OF RISK

The word risk can conjure up many thoughts in people's minds, depending on the point of view of the listener. A project manager may think of risk to on-time completion of the project. A CEO may think of the business risk. An engineer may think of risk of infringing on someone else's intellectual property.

Examples of types of risk:

• Project risk	• Technical risk
• Financial risk	• Schedule risk
• IP risk	• Regulatory risk
• Security risk	• Safety risk

In this book, we focus only on safety risk. It's important that when you use the words "risk management" to make sure your audience understands which type of risk you are talking about.

Due to inadequate understanding of risk, *harm* and *risk* are frequently confused. The manufacturers are required to disclose residual *risks* of medical devices, or pharmaceuticals. As seen in the example of Lipitor in Section 4.1, typically a list of *Harms* is presented to fulfill this requirement. This book helps to provide more clarity and distinctions on these terms to help reduce or prevent such confusions.

4.3 CONTRIBUTORS TO RISK

A major contributor to risk is the human, be it the designer, builder, maintainer, or the user of the System. Assuming good intention, the main cause of human influence on risk is inconsistency between reality, and the mental model of reality that the human holds. Mental models are necessary to function in life. Model: putting your hand in boiling water will cause a burn. Imagine if you lost this mental model and had to relearn this every time.

Leveson [12] says all models are abstractions; they simplify the thing being modeled by abstracting away what are assumed to be irrelevant details and focusing on the features of the phenomenon that are judged to be the most relevant. Selecting some factors as relevant and others as irrelevant is, in most cases, arbitrary and entirely the choice of the modeler. That choice, however, is critical in determining the usefulness and accuracy of the model in predicting future events.

The mental models that we hold are hypotheses based on theories or empirical observations. Their usefulness depends on how accurately they model reality. If models that we employ in risk management are incorrect, the safety Risk Controls that we devise may not be effective in reducing risk.

The event-chain model of causation is one of the most common ways of modeling accidents. This was depicted in Fig. 3.1. Extending the event-chain model to risk management, suggests that the most obvious countermeasure for preventing Harm is to break the chain of events before the Harm happens. While this is a useful model, care should be taken to consider factors that could indirectly affect the model. For example, in the 1970s if you wanted to safely stop a car while driving on ice, the advice was to pump the brake pedal, so as to avoid locking the wheels and skidding. Modern cars are equipped with antilock brakes. When driving a car with antilock brakes, the way to stop the car most quickly is to step on the brake pedal as hard as you can, because the brake system does the pumping automatically. Using the model of braking from older cars, on a modern car, actually reduces the braking capability of the car and increases the stopping distance, hence increasing the risk of a car crash.

4.4 RISK PERCEPTION

ISO 14971 [3,7] says "It is well established that the perception of risk often differs from empirically determined risk estimates. Therefore, the perception of risk from a wide cross section of stakeholders should be taken into account when deciding what risk is acceptable. To meet the expectations of public opinion, it might be necessary to give additional weighting to some risks. In some cases, the only option could be to consider that identified stakeholder concerns reflect the values of society and that these concerns have been taken into account."

Risk perception and tolerance are strongly influenced by human psychology. The same circumstance would be perceived and tolerated differently by different people. In fact, this is why stock markets work—some people think that the risk of losing money is high, so they sell, while others think the risk of losing money is low and they buy.

When an adverse event happens, the public perception of the risk of that event suddenly jumps up. In the early 1980s Aeroflot, the Russian airlines, had a string of airplane crashes. All the airplanes involved in the crashes were Tupolev model Tu-154. At that time, there were people who would refuse to get on their flight if the brand of aircraft for the flight was Tupolev. We see this in any industry, including the medical device industry. When a particular medical device is involved in an adverse event, all other devices of the same type become suspect. You may remember adverse events related to silicone gel breast-implants ruptures, that terrified many women, even if nothing happened to them.

Risk perception is also influenced by other factors. For example, whether exposure to the Hazard seems to be involuntary, avoidable, from a man-made source, due to negligence, arising from poorly understood causes, or directed at a vulnerable group within society. People tend to be more tolerant of natural risks than risks due to man-made sources. Risk to children is less tolerated than risk to adults.

Michael Lewis says in the Undoing Project [13] "What people call risk aversion is tantamount to a fee that people pay, willingly, to avoid regret – a regret premium." Let's say you are at a horse race, and you have $5 to bet. They give you 10 to 1 odds on a horse. That is, there is a chance that in a short amount of time you can get 1000% return on your money. Or, you can lose your $5. Avoiding taking this risk means you are willing to pay a $50 premium not to feel the regret of losing $5, if your horse didn't win. How this plays out in medical devices is that people are willing to forego a high chance of benefit from a device, to avoid a low chance of receiving health damage from the device. This calculus changes if a person is desperate. For example, terminal patients are willing to accept higher risk devices for a lower benefit.

4.5 RISK COMPUTATION

Clause 4.4 of ISO 14971 [3,7] requires that the manufacturers estimate the risks associated with each Hazardous Situation. The Standard [3,7] offers a figure in its annex E, which is replicated in Fig. 4.1. The interpretation of this figure is that after a Hazard has manifested, through the progression of a sequence of events, a Hazardous Situation is realized, where people, property, or the environment are exposed to the Hazard(s). The probability of occurrence of the Hazardous Situation is called P_1. Given the Hazardous Situation, there is a probability that the subject

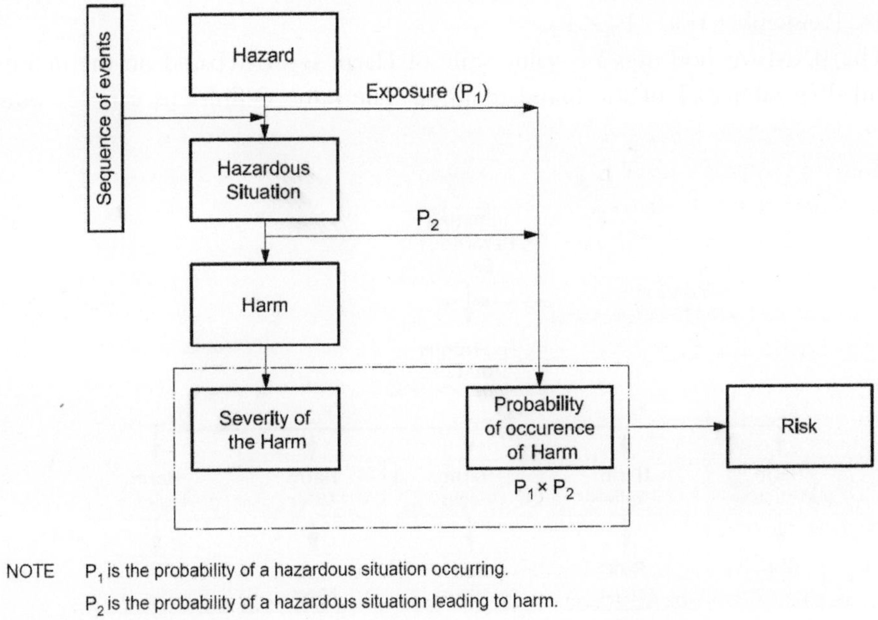

Figure 4.1 ISO 14971 Figure E.1.

(people/property/the environment) can be harmed. This probability is called P_2. The product of P_1 and P_2 is risk. Another way to interpret this is as follows.

Risk is the probability of sustaining Harm in a Hazardous Situation

Notice that in Fig. 4.1 there is a *dotted line* around the [Severity of the Harm] and [Probability of occurrence of Harm]. The reason for this is that given the same Hazardous Situation, different people (subjects) experience different degrees of harm. For example, when exposed to the flu virus, some people with a strong immune system may experience nothing, while others may show the typical symptoms of fever and aches, and some people might even die from the flu. Without prescribing how to address these different degrees of harm, the Standard [3,7] leaves the door open for the manufacturer to choose how to address this phenomenon.

Traditionally, many manufacturers take the conservative route and assume the worst-case Harm. This creates two problems. First, for many Harms there is a chance, though small, that the patient may die. Manufacturers who use this strategy find themselves facing an exaggerated picture of the hazards of their device—a picture that seemingly shows patients could die due to most of the Hazards related to the device, even though historical data shows otherwise. Those manufacturers find themselves forced to overdesign their devices, wasting a lot of engineering resources. The second problem is that resources are then spent on the highest-severity Harms based on worst-case analysis, while moderate severity Harms that are actually more probable, get less attention. This means a lower risk Harm would get priority over a higher risk Harm. Remember risk $= P_1 \times P_2$.

The BXM method uses a 5-value scale of Harm severity based on the definitions provided in table D.3 of the Standard [3,7]. The same definitions can be found in

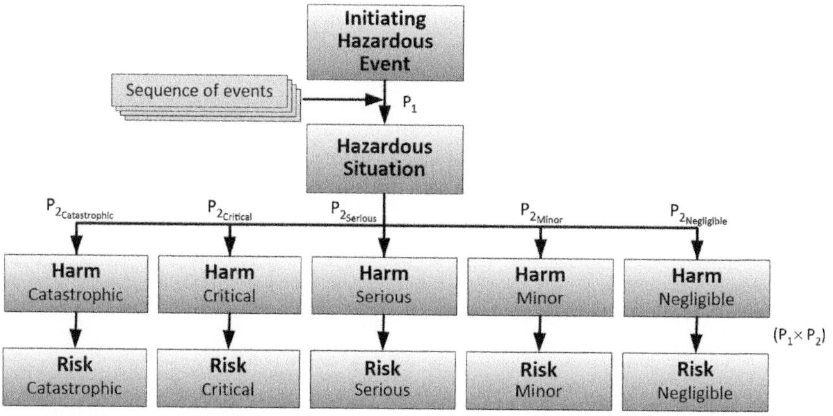

Figure 4.2 5-scale risk estimation.

Table 17.3 in Section 17.2. With the 5-value Harm severity model, an enhancement to Figure E.1 of the Standard [3,7] can be made (see Fig. 4.2).

Note that P_1 has dimension, but P_2 is dimensionless. Therefore, Risk, which is $P_1 \times P_2$, has the same dimensions as P_1.

Some P_1 examples:

- Probability of exposure of the patient to a damaged cannula is 1 in 1000 uses.
- Probability of a patient not receiving pacing from their pacemaker is $<10^{-5}$ per patient-year.

Examples of P_2:

- Probability of death from intracranial hemorrhage is 2.5%.
- Probability of permanent or life-threatening injury from electrocution is 4%.

CHAPTER 5

Risk Management Standards

Abstract

There are a number of standards that address safety for medical devices. For example, ISO 14971, IEC 60601-1, IEC 62304, IEC 62366, ISO 10993-1, and so on. ISO 14971 is the central standard for risk management of medical devices and is recognized both in the EU and the United States. Compliance with ISO 14971 is the most common way of establishing the case for the safety of a medical device.

Keywords: ISO 14971; IEC 60601-1; EN ISO 14971:2012

There are a number of standards that address safety for medical devices. For example, ISO 14971 [3,7], IEC 60601-1 [8], IEC 62304 [9], IEC 62366 [14], ISO 10993-1 [15], and so on (see Fig. 5.1). ISO 14971 is the central standard for risk management of medical devices and is recognized both in the EU and the United States. Other medical device safety standards make normative references to ISO 14971 [3,7]. ISO 14971 [3,7] defines a set of requirements for performing risk management, but does not stipulate any specific process.

Most of the safety standards are becoming progressively more outcome-based and less prescriptive. They specify what outputs are expected, and give you the freedom to choose your own methods. This is both a blessing and a curse. On the one hand, you get freedom to create your own process. On the other hand, the standards don't tell you how to do things, so the method that you choose becomes subject to questioning and you must be prepared to defend it.

For all regulated medical devices, the manufacturers must show that formal risk management processes have been applied and that their respective safety risks have been reduced to acceptable levels.

5.1 ISO 14971 HISTORY AND ORIGINS

ISO 14971 was originally developed in 1998 with participation from 112 countries. It is widely recognized as the official international standard for medical device risk management.

To produce ISO 14971, ISO/TC 210 WG4 was formed. Simultaneously IEC 60601-1 3rd edition was planning to include risk management for medical devices. In

DOI: https://doi.org/10.1016/B978-0-12-813098-8.00005-2

© 2018 Elsevier Ltd.
All rights reserved.

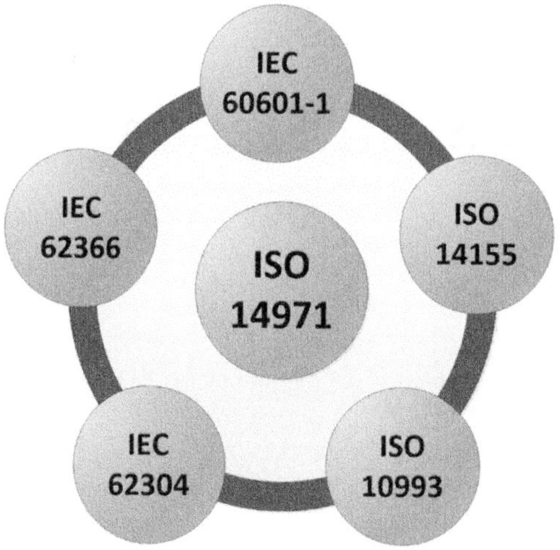

Figure 5.1 ISO 14971, a central standard.

order to combine efforts, JWG1 IEC/SC 62A was formed, so 14971 is the product of work by both IEC and ISO. The ISO designation was chosen because 14971 is not just about medical *electrical* equipment, but about all medical devices.

ISO 14971 was first released in the year 2000. The second edition was promulgated in March 2007, and it canceled and replaced the 2000 version. ISO 14971:2007 [3] is still current in the United States. European harmonized standard EN ISO 14971:2012 [7] identifies discrepancies between ISO 14971:2007 [3], and EU directives: 93/42/EEC on Medical Devices [4], 90/385/EEC on AIMDs, and 98/79/EC on in Vitro Diagnostics, and adds three annexes: ZA, ZB and ZC to ISO 14971:2007 [3].

Prior to ISO 14971 there was no other standard to address the risk management of medical devices. ISO 14971 was drafted as a *framework* because there are so many different types of medical devices, that it is difficult to produce a specific process that would cover all devices. ISO 14971 is applicable to the complete product lifecycle, from concept to decommissioning and disposal.

5.2 HARMONIZED STANDARDS

In a strict legal sense, regulatory entities do not "require" compliance with either ISO 14971:2007 [3], or EN ISO 14971:2012 [7]. Under EU legislation, the use of standards (or harmonized standards) is purely voluntary and you are free to comply with the Essential Requirements of the Directives via other means. However, with regard to risk management, ISO 14971 is the recognized reference standard, and compliance to it makes it easier to persuade a notified body that your device is acceptably safe.

CHAPTER 6

Requirements of the Risk Management Process

Abstract

ISO 14971 offers a framework for managing the risks of medical devices. Rather than being prescriptive, this framework has specific requirements and expectations.

Keywords: Risk management process

ISO 14971 [3,7] offers a framework for managing the risks of medical devices. Rather than being prescriptive, this framework has specific requirements and expectations. Namely:

- Have a documented process, and apply it to the entire product lifecycle
- Have a plan
- Use qualified personnel to perform risk management
- Ensure completeness of Risk Controls for all identified risks
- Evaluate the overall residual risk
- Show that the benefits outweigh the risks
- Monitor production and postproduction information and feed that back into the risk management process
- Produce a risk management report
- Create and maintain a risk management file
- Show traceability among Hazards, risks, risk analysis, evaluation and control, and verification of Risk Controls

The manufacturer is free to create the internal processes to meet the above requirements.

6.1 RISK MANAGEMENT PROCESS

The risk management process should have at least the following element:

6.1.1 Risk analysis

The requirements for risk analysis are:

- Provide a description and identification of the medical device under analysis
- Define the scope of analysis—what's included and what's excluded

Safety Risk Management for Medical Devices
DOI: https://doi.org/10.1016/B978-0-12-813098-8.00006-4
© 2018 Elsevier Ltd.
All rights reserved.

- Identify the intended use of the medical device
- Identify the safety characteristics of the medical device
- Identify the Hazards that the device could present
- Estimate the risk for each Hazard

You are required to identify the persons who perform the risk analysis. This is necessary for confirmation of qualification of such individuals. Also, the date of the analysis must be recorded.

6.1.1.1 Hazard identification

The Standard [3,7] requires documentation of both known, and foreseeable Hazards associated with the medical device under both normal and fault conditions. This creates four types of Hazards. Table 6.1 captures this Hazard taxonomy.

A number of techniques are used to identify the System Hazards. For example, Fault Tree Analysis (see Section 12.1 for details), and Failure Modes and Effects Analysis (see Section 12.4 for details).

Other techniques of Hazard identification are: search of published literature for reported Hazards of similar devices; search of databases such as MAUDE [16], and EUDAMED [17]; and examination of ISO 14971 [3,7] Annex E, Table E.1. In Table E.1, there is a listing of many common Hazards that could help with the identification of Hazards that are related to your medical device.

Table 6.1 Hazard taxonomy

		Hazard Type	
		Known	**Foreseeable**
Fault Condition	Normal	X	X
	Fault	X	X

6.1.1.2 Risk estimation

Risk estimation methods are explained in detail in Chapter 17, Risk Estimation.

6.1.2 Risk evaluation

Risk evaluation requires knowledge of risk acceptance criteria. See Chapter 19, Risk Evaluation, for a detailed discussion of risk evaluation.

6.1.3 Risk controls

See Chapter 18 for a detailed discussion of Risk Controls.

6.1.4 Risk control verification

See Chapter 22, Verification of Risk Controls, for a detailed discussion of verification of Risk Controls.

6.1.5 Monitoring

Once the risk management process is completed and the product is released to the market, it is required that manufacturers continuously monitor for how the device is performing. This can take the form of surveillance, which is active seeking of information. Or, complaint handling, which is the passive method of receiving information. All collected information is to be used to update the risk management work products as necessary. See Chapter 24, Production and Postproduction Monitoring, for a detailed discussion of production and post-production monitoring.

CHAPTER 7
Quality Management System

Abstract

Your Quality Management System (QMS) is the internal reflection of the external standards. Your SOPs bring the standards to life within your company. A clear, organized, and well-written SOP sends a message to regulatory bodies that your QMS is consistent and compliant to applicable standards. More importantly, well-written SOPs ensure that your personnel can faithfully and accurately follow your internal processes resulting in proper outcomes and quality work. On the other hand, a poorly written confusing SOP could result in noncompliance, which could lead to observations and warnings from regulatory bodies at best and in some cases, severe penalties including consent decrees and shut down of the company.

Keywords: Quality Management System; risk management; SOP

Your Quality Management System (QMS) is the internal reflection of the external standards. Your SOPs bring the standards to life within your company. A clear, organized, and well-written SOP sends a message to regulatory bodies that your QMS is consistent and compliant to applicable standards. More importantly, well-written SOPs ensure that your personnel can faithfully and accurately follow your internal processes resulting in proper outcomes and quality work. On the other hand, a poorly written confusing SOP could result in noncompliance, which could lead to observations and warnings from regulatory bodies at best and in some cases, severe penalties including consent decrees and shut down of the company.

In addition to SOPs, many companies employ templates and work instructions to assist in the quality execution of the risk management work. While SOPs are more high level and designed to ensure compliance to applicable standards, work instructions provide detailed guidance on how to do the work.

As with all other controlled documents, risk management artifacts must be controlled per the methods that are stipulated in your QMS. Absent this strict control, it would be easy for analyses and their targets to go out of sync. That in turn could lead into unwanted outcomes like missed Hazards, injured patient, audit findings, etc.

One of the great benefits of a good risk management process is the ability to provide safety impact analysis of proposed changes. In any good QMS, an impact

Safety Risk Management for Medical Devices
DOI: https://doi.org/10.1016/B978-0-12-813098-8.00007-6
© 2018 Elsevier Ltd.
All rights reserved.

analysis is done for any proposed change to a controlled design. One aspect of this impact analysis should be the determination of the effect of the proposed change on safety. Without the benefit of a formal risk management process, estimation of the safety impact of a proposed change would be just a guess.

CHAPTER 8

Usability Engineering and Risk Analysis

Abstract

Medical devices are providing ever more benefits and at the same time are becoming more complex with user interfaces that are not so intuitive, difficult to learn, and difficult to use. Use-errors caused by inadequate medical device usability have become an increasing source of adverse safety events. As healthcare evolves, and more home-use is promoted as a cost-saving measure, less skilled users including patients themselves are now using medical devices. This chapter examines the contribution of usability engineering to the safety of medical devices.

Keywords: Use Failure; Use Error; usability engineering; IEC 62366; human factors; user-device interaction; task analysis

Medical devices are providing ever more benefits and at the same time are becoming more complex with user interfaces (UIs) that are not so intuitive, difficult to learn, and difficult to use. Use-errors caused by inadequate medical device usability have become an increasing source of adverse safety events. As healthcare evolves, and more home-use is promoted as a cost-saving measure, less skilled users including patients themselves are now using medical devices.

In this chapter, the terms "Use Failure" and "Use Error" are used interchangeably.

The leading standard for usability engineering is IEC 62366, which in its latest edition splits into two documents: IEC 62366-1:2015 [14], and IEC TR 62366-2:2016. This international standard describes a usability engineering process to provide acceptable safety risks as related to the usability of a medical device. It should be noted that the scope of usability engineering is larger than safety. Usability engineering is also concerned with how well and easily the user can interact with the medical device in order to achieve their desired outcomes. Therefore customer experience and satisfaction are also influenced by usability engineering.

IEC 62366 [14] provides a process to assess and mitigate risks associated with the correct-use and use-errors. Malice, and Abnormal use are excluded from the risk management process. See Sections 8.1 and 8.2 for further elaboration and elucidation of these terms.

It should be noted that the FDA considers *Human Factors* and *Usability Engineering* to be synonymous. See FDA Guidance on HFE [18], par 3.6. Also, the FDA has replaced the term "user error" with "use error." This means that use error is considered by the FDA to be a *device* nonconformity because human factors should be considered in the design process.

Safety Risk Management for Medical Devices
DOI: https://doi.org/10.1016/B978-0-12-813098-8.00008-8
© 2018 Elsevier Ltd.
All rights reserved.

Tip Start usability engineering right from the start of product development. Though FDA requires summative studies, they also want a summary of how usability engineering was performed throughout the design process.

8.1 KEY TERMS

Below, a select group of terms are defined. A clear understanding of these terms is important in the proper dispositioning of use-errors, and is also beneficial for communication with team members.

Term	Definition
Abnormal Use	Conscious, intentional act or unintentional act that is counter to or violates normal use and is beyond further reasonable means of user-interface-related risk control by the manufacturer [14] 3.1
Action Error	One of the causes of Use Error. Related to error in performance of an action, and not primarily due to perception or cognition deficiencies
Cognition Error	One of the causes of Use Error. Related to deficiency in cognition, and not primarily due to perception deficiencies or the inability to perform the action
Correct Use	Normal Use without Use Error [14] 3.3
Effectiveness	Accuracy and completeness with which users achieve specified goals [14] 3.4
Formative Evaluation	User interface evaluation conducted with the intent to explore user interface design strengths, weaknesses, and unanticipated use errors [14] 3.7 Also known as Formative Study
Hazard-Related Use Scenario	Use scenario that could lead to a hazardous situation or harm [14] 3.8
Lapse	A memory failure. The User had the knowledge but forgets the knowledge and makes a decision based on incorrect knowledge. User executes the decision without slips This term was used in IEC 62366 2007 version, but not used in the 2015 version
Malice	Intentional act to do harm to people, property, or the environment
Mistake	The User does not have the knowledge and/or User makes a wrong decision. User executes without slips This term was used in IEC 62366 2007 version, but not used in the 2015 version

(Continued)

(Continued)

Term	Definition
Misuse	Incorrect or improper use of the medical device [3] 4.2 *Note*: This is not an official definition
Normal Use	Operation, including routine inspection... according to the instructions for use or in accordance with generally accepted practice for those medical devices without instructions for use [14] 3.9
Perception Error	One of the causes of Use Error. Related to deficiency in perception, and not primarily due to cognition deficiencies or the inability to perform the action
Primary Operating Function	Function that involves user interaction that is related to the safety of the medical device [14] 3.11
Slip	The Use has the knowledge and the intention, but executes incorrectly. Attentional error This term was used in IEC 62366 2007 version, but not used in the 2015 version
Summative Evaluation	User interface evaluation conducted at the end of the user interface development with the intent to obtain objective evidence that the user interface can be used safely [14] 3.13 *Note*: Summative evaluations are frequently used to provide verification of effectiveness, of risk controls
Task	One or more user interactions with a medical device to achieve a desired result [14] 3.14 *Note*: A task can be further broken down into "steps"
Usability	Characteristic of the user interface that facilitates use and thereby establishes effectiveness, efficiency, and user satisfaction in the intended use environment [14] 3.16
Use Error	User action or lack of user action while using the medical device that leads to a different result than that intended by the manufacturer or expected by the user [14] 3.21 *Note*: Use Error includes the inability of the user to complete a task
Use Failure	Failure of a User to achieve the intended and expected outcome from the interaction with the medical device *Note*: Inability to perform a task, as well as Use Errors are Use Failures
Use Scenario	Specific sequence of tasks performed by a specific user in a specific use environment and any resulting response of the medical device [14] 3.22
User Interface (UI)	Means by which the user and medical device interact [14] 3.26 *Note*: User interface includes all the elements of the medical device with which the user interacts including the physical aspects of the medical device as well as visual, auditory, tactile displays and is not limited to a software interface

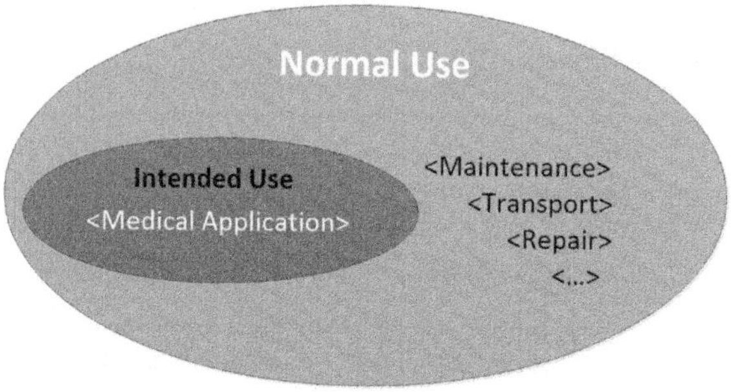

Figure 8.1 Type of normal use.

8.2 DISTINCTIONS

As defined in Section 8.1, Abnormal Use is: "Conscious, intentional act or unintentional act that is counter to, or violates normal use and is beyond further reasonable means of user-interface-related risk control by the manufacturer" [14]. Examples include: reckless use, sabotage, or intentional disregard for safety instructions.

Malice is intentional act with the aim of causing harm. The distinction between Malice and Abnormal Use is that Malice has the intention to do harm, while Abnormal Use is recklessness that *might* cause harm.

Setting aside Abnormal Use and Malice, we are left with Normal Use. Normal Use includes the Intended Use of the device, i.e., the medical purpose of the device, as well as other ancillary uses such as maintenance, transport, etc. (see Fig. 8.1). Also see IEC 62366 [14] Note 1 to Section 3.9 for further elaboration on this concept.

Any Normal Use is subject to Use Errors, which could lead into a Hazardous Situation.

8.3 USER-DEVICE INTERACTION MODEL

Modeling of human interaction with technological systems helps with prediction of use errors and has its roots in aerospace and defense for the design of aircraft cockpits and crew stations. They model the human in three parts:

1. input/sensing
2. processing
3. output

Fig. 8.2 shows a model of the User-Medical Device interaction. In this figure, there are two possible ways in which a Use Error can lead into a Hazardous Situation.

1. The User enters an erroneous input into the Medical Device, which in turn produces an output that is hazardous, e.g., User enters the wrong dose on the UI of an infusion pump.

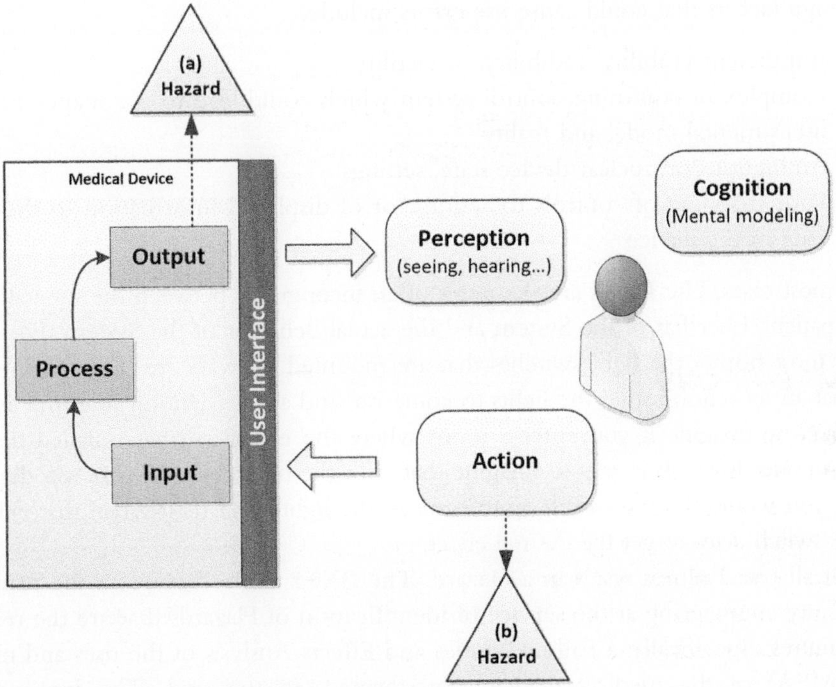

Figure 8.2 Model of User-Medical Device interaction.

2. Through either a perception error, or a cognition error, the User takes an action that creates a Hazard. This action could be on the Medical device, or outside of the Medical Device, e.g., a diabetic patient misreads, or misinterprets the glucose reading on a glucose monitor and administers too high an insulin dose to him/herself.

Using this model facilitates prediction of potential use errors, and also helps identify the required human capabilities for interacting with the medical device. This in turn can guide the design of the UI for safer and more effective medical devices.

Perception examples: See, Hear, Feel. Perception may be distorted, disrupted, or impossible depending on the environmental conditions. For example, a surgeon may not be able to feel surface texture with a gloved hand.

Cognition examples: Interpret, Know, Compute, Decide. Cognition also, may be difficult or compromised under conditions of stress, fatigue, heavy or multitasking workload.

Action examples: Press, Touch, Twist, Pull, Push, Follow. Actions may be challenged under conditions of fatigue, pain, injury, or barriers such as gloved operations.

Design factors that could cause use errors include:

- Insufficient visibility, audibility, or tactility
- Complex or confusing control system which could lead to dissonance between user's mental model and reality
- Ambiguous or unclear device state, settings
- Poor mapping of controls to actions, or of displayed information to the actual state of the device

In most cases, Use Errors are the result of an incongruity between the mental model that a patient/User has of the System and the actual behavior of the System. For example, in most houses the light switches that are mounted on walls are oriented in such a way that an *up* action causes the lights to come on, and a *down* action causes the lights to go off. Now imagine if you enter a room where the electrician has installed the light switch upside-down. It is easy to imagine that with the mental model that was described earlier, you would flip the switch *up* to turn on the light. And then when that fails, you flip the switch *down* to get the desired effect.

Not all Use Failures result in a Hazard. The BXM method employs the principles of usability engineering at the service of identification of Hazards that are the result of Use Failures. Specifically, a Failure Modes and Effects Anslysis of the uses and misuses (UMFMEA) of the medical device is performed, to this end. The input to the UMFMEA is the medical device use-scenarios and task analyses. See Section 12.8 for further details on UMFMEA.

8.4 USE FAILURES

In this section, we delve more deeply in Use Failures. As defined in Section 8.1, a Use Failure is the failure of a User to achieve the intended and expected outcome from the interaction with the medical device. How can this happen?

1. The user is unable to perform the task.
   This could be due to confusing UI, physically not doable action by the User, etc.
2. The User completes the task but with significant difficulty.
   For example, due to a confusing UI, the user struggles, makes error(s), but notices the error(s) and corrects it, and in the end completes the task. Although by the strict definition of Use Failure, this would not be a Use Failure, because the user eventually completes the task, it is advisable to count this type of interaction as Use Failure so that it receives the appropriate attention.
   Close-calls, or near-misses where a user almost fails a task, should also be treated as Use Failures so that they could inform design decisions for improvements.

3. The User executes the task erroneously. Reasons:
   - Perception error
     Example: User cannot read a display due to: font size, contrast, brightness, layout, etc.
   - Cognition error
     - Memory failure
       - Inability to recall knowledge which was gained before.
       - Lapse. User has the knowledge but temporarily forgets, or omits an action.
     - Rule-based failure
       - Incorrect application of a rule to the task.
       - Application of an inappropriate rule to the task.
     - Knowledge-based failure
       - User does not have the knowledge, e.g., due to no training, or not reading the instructions for use.
       - User applies the wrong mental model, e.g., improvisation under unusual circumstances
     - Attentional error
       - Slip. User has the knowledge and the intention, but executes incorrectly, e.g., transposing two keys on a keyboard.
       - Physical Error, e.g., pressing the wrong button; reason: buttons are too small and too close to each other. Or, double hitting of a button on a UI, e.g., intending to program a dose on an infusion pump by pressing 4, but accidentally pressing 44 and getting the wrong dose.

8.5 ENVIRONMENTAL FACTORS

Environmental factors are a contributing Cause to Use Failures. Some environmental factors that could play a role in causing use failure are:

- Temperature
- Humidity
- Vibration
- Atmospheric conditions
- Distractions
- Lighting
- Physical surroundings
- Acoustic noise
- Workload-related stress
- Other systems and devices in the environment of use

While you may not have any control over the environmental factors that surround the medical device, knowledge and awareness of such factors is critical for the proper design of the UI of the medical device.

Tip Consider the typical workflows where your device will be used. If your device replaces an existing device, legacy behaviors may carry over to the use of your System.

8.6 DESIGN MEANS TO CONTROL USABILITY RISKS

In most cases, changes to the design of the medical device can serve to reduce the risks due to use failures. Some examples of such design controls are as follows:

- *Keystroke debouncing*—if the same key is pressed within 200 ms, ignore the second keystroke.
- *Reasonableness checks*—evaluate the user input for reasonableness. If the input is out of range or unreasonable, inform the user.
- *Proper sizing*—use anthropomorphic data to size the use interface such that physical errors are less likely, e.g., buttons sized to human fingers.
- *Alarm types*—use IEC 60601-1-8 [19] for guidance on proper design of alarms.
- *Font size*—use AAMI HE75 [20] for guidance on font sizes for visual displays.

If such design means are employed to reduce risk, they should be verified for effectiveness in risk reduction.

8.7 TASK ANALYSIS

The Standard [14] requires that the manufacturers identify UI characteristics that could be related to the safety of medical devices. A common tool to achieve this is task analysis. Task analysis is a formal and systematic activity that starts by creating a detailed description of sequential and simultaneous actions of the user of the medical device. Task analysis usually starts with high-level use-scenarios, and then adds tasks and ultimately details down to individual steps are spelled out. The output of task analysis is input to the UMFMEA (see Section 12.8). As such, conditions of use and user profiles are important to know, and should be included in the task analysis. Task analysis results are typically stored in a tabular or flowchart format.

Task analysis should begin at concept development stage at a high level, and should progress with more details as UI design matures. Because designs iterate multiple times during the design process, task analysis should be kept in sync with the UI design to ensure the validity of risk management with respect to UI design.

For the purposes of risk management, what is of interest are the user-performed steps, and how errors in performing them could result in Hazards.

8.8 USABILITY AND RISK

Some Use Failures can result in Hazards. The UMFMEA analyzes all Use Failures, and captures the Hazards that are due to Use Failures in the End Effects column. Just as in other FMEAs, the End Effects which are Hazards are then captured in the Risk Assessment and Control Table for risk estimation.

Risk is the product of P_1, the probability of occurrence of a Hazardous Situation and P_2, the probability of experiencing Harm from the Hazardous Situation. P_1 itself is the product of *P*(Hazard) and *P*(Exposure to the Hazard). Ordinarily, *P*(Hazard) can be derived from the Occurrence rating in FMEAs. However, with respect to risk, IEC 62366 [14] Annex A, Subclause 5.5 makes the following statements:

"Selection of the HAZARD-RELATED USE SCENARIOS can be based on the SEVERITY of the potential consequences of the associated HAZARDS. It can be needed in this way to focus on HAZARDS rather than RISKS because the probability of occurrence of encountering a HAZARD, which is one component of RISK, can be very difficult to estimate, especially for a novel MEDICAL DEVICE for which no POST-PRODUCTION data are available.

Another basis for selection of the HAZARD-RELATED USE SCENARIOS is the RISK of the occurrence of HARM to the PATIENT or USER. These values can also be difficult to determine, as they are based on assumptions closely related to probability of occurrence and without data, can be difficult to justify. Finally, and only in the presence of data that provides a justification, should RISK values based on the combination of SEVERITY and probability of occurrence of the HAZARD be used as the basis for prioritization of HAZARD-RELATED USE SCENARIOS. Values for these probabilities or probability of occurrence can be derived from POST-PRODUCTION data on current or previous versions of the same MEDICAL DEVICE or on the level of certainty that the RISK CONTROL measures are effective, which should also be justified with data."

In other words, the Standard [14] suggests that unless you have data, you should presume the probability of exposure to Hazards due to Use Failures is 1, and judge the use scenarios based on the severity of associated Harms.

ISO 14971 [3,7] provides some guidance in section D.3.2.3 that when estimate of P_1 cannot be made on the basis of accurate and reliable data, or when a reasonable qualitative estimate is not possible it is best to set $P_1 = 1$ and focus on using Risk Control measures to eliminate the Hazard, or reduce the probability of Harm, or reduce the severity of the Harm.

Naturally, elimination of a Hazard is best, but not always possible. The manufacturer has two options:

1. Gather data, upon which P_1 could be estimated, and thus risk could be computed
2. Assume $P_1 = 1$, and use Risk Controls to reduce the probability occurrence of Use Failure as far as possible, and add controls to reduce the severity of the Harm

In the following subsections, each option is explored.

8.8.1 Data gathering

If the medical device is based on an existing released product, for which postproduction data is available, derive probability of occurrence of Hazardous Situations due to Use Failures from the available data. Use this probability in conjunction with P_2 data to estimate the risks due to Use Failures.

If the medical product is new, or the part of UI that is under analysis is new, or if postproduction data of sufficient quality is unavailable, then plan and execute formative and summative studies to generate the necessary data to support the P_1 estimates. Compute the risks of Use Failures based on the P_1 data.

8.8.2 Risk reduction and compliance with IEC 62366 process

Absent a value for P_1, risk of Use-Failures cannot be estimated. The alternative is to use the process that is prescribed by IEC 62366 [14] to reduce Use Failures to the degree possible. Additionally, control measures could be implemented to reduce the severity of the Harm, should the Harm happen.

As Use-Failure risks are not estimated, they cannot be included in the overall residual risk computations.

CHAPTER 9

Biocompatibility and Risk Management

Abstract

The International Standard ISO 10993-1 is intended for the protection of humans from potential biological risks from the use of medical devices. It describes the process for biological evaluation of medical devices within the framework of a risk management process. ISO 10993-1 describes biological evaluation of medical devices as a design verification activity in the broader context of risk management, which is governed by ISO 14971.

Keywords: Biocompatibility; biological testing; ISO 10993; biological hazards

The International Standard ISO 10993-1 [15] is intended for the protection of humans from potential biological risks from the use of medical devices. It describes the process for biological evaluation of medical devices within the framework of a risk management process.

Annex B of the Standard [15], describes biological evaluation of medical devices as a design verification activity in the broader context of risk management, which is governed by ISO 14971 [3,7]. Annex B includes guidance on the application of ISO 14971 [3,7] to the conduct of biological evaluation.

Sections B.2 and B.3 of Annex B of ISO 10993-1 [15] describe "a continuous process by which a manufacturer can identify the biological hazards associated with medical devices, estimate and evaluate the risks, control these risks, and monitor the effectiveness of the control."

In general manufacturers aim to use existing materials that are proven to be acceptably safe for medical use from a biological perspective. However, at times it could be found that existing proven materials are not suitable for the specific medical device at hand.

When selecting a material for a medical device which would come in contact with the patient, particularly for implantable devices, it is important that all Hazards due to the use of the selected materials be identified, and the risks for each Hazard be estimated and evaluated.

Application of ISO 14971 [3] risk management methodology can identify, estimate, and evaluate the risks due to biological hazards and judge their acceptability.

Safety Risk Management for Medical Devices
DOI: https://doi.org/10.1016/B978-0-12-813098-8.00009-X
© 2018 Elsevier Ltd.
All rights reserved.

As biological evaluation is a component of medical device risk management activities, it must be planned in advance. This includes: literature searches, biological evaluations, review and approvals of biological evaluations, and documentation of residual risks.

In biological evaluations, the safety risk to patient/user is a function of the toxicity of the materials, their route and duration of exposure, and the physical properties of the materials. For example, a rough surface creates a larger exposure area than a smooth surface.

ISO 14971 [3,7] and ISO 10993 [15] require characterization of medical devices for potential Hazards. This includes the materials themselves, all additives and processing aids, interaction with sterilization processes, and chemical transformation of the materials (e.g., degradation) during use. Also consider the interaction with, or contamination from the packaging materials. Both direct and indirect tissue contact of medical device materials should be considered.

Factors that can influence the biological safety of medical device materials include:

- Mechanical wear, strain, vibration, and deformation due to use
- Biomechanical interactions such as abrasion, friction, and sticking
- Biochemical interactions such as with acids, and enzymes
- Heat/cold
- Radiation
- Chemicals, such as ethylene oxide in the course of sterilization
- Cleaning solutions

Risk estimation requires two components: (1) the likelihood of exposure to a Hazard, and (2) the probability of sustaining Harm from the Hazardous Situation.

Determination of the likelihood of exposure to a biological hazard depends on factors including:

a. The bioavailability of toxic materials—how likely is it for the toxin to become present?
b. The potential for exposure

Determination of the probability of sustaining Harm in the event of exposure depends on factors including:

a. The nature of the toxin
b. The quantity of the toxin (dose response)
c. The tissue that is exposed (e.g., intact skin vs mucous membranes)
d. The duration of exposure
e. Geometric properties (e.g., particle size, porosity, surface texture)

Much information can be derived from published literature, existing in-house data, or suppliers of materials. Where such information is unavailable, chemical or physical characterization, or biological testing may be required to gather the required data.

Data requirements are less stringent for lower risk applications, e.g., temporary intact skin contact, than for higher risk applications, e.g., brain implants.

Once the risk has been estimated, it is evaluated against the risk acceptance criteria, which are defined in the risk management plan. For reduction of risks, certain measures can be taken to bring the risks down as far as possible. For example, replacement of a material that has toxicological risks, with a material that doesn't have toxicological risks is a design that is inherently safe. Other potential Risk Control measure:

- Reducing exposure time
- Reducing exposure surface area
- Use of coatings/materials that reduce adverse biological response
- Changes to manufacturing processes to reduce/eliminate toxic additives, or manufacturing aids
- Better cleaning/rinsing processes to remove toxic residues

Keep in mind that some Risk Control measures may introduce new Hazards, or increased risks elsewhere in the design. In such cases, some retesting may become necessary.

Biological evaluation of medical device materials relies on risk assessment to provide justification for not conducting certain testing. This is valuable from a project cost and schedule perspective, and more importantly from the ethical perspective in that some animal testing may be avoided.

An important factor to consider is that even if a given material is shown to be sufficiently safe by itself, it cannot be deduced that the same material when used in combination with other materials will still be safe. Therefore the total device in its final form, produced using the final processes is typically subject to biological testing.

As described in Chapter 24, Production and Postproduction Monitoring, data from production and postproduction must be monitored for any occurrences of adverse effects, including due to biocompatibility. Such learnings must be used to update the risk management file as necessary.

CHAPTER 10

The BXM Method

Abstract

The methodology presented in this book is called BXM. The BXM method is compliant with ISO 14971. The basic premise of the BXM method is to decompose the System; do risk analysis on the parts of the System; and then integrate the underlying analyses into the System-level analysis. This offers the ability to use parallel work streams, optimize labor and skills usage, and reuse analyses.

Keywords: BXM method; quantitative risk estimation; software-assisted automation

The methodology presented in this book is called BXM. The BXM method is compliant with ISO 14971 [3,7]. The basic premise of the BXM method is to decompose the System; do risk analysis on the parts of the System; and then integrate the underlying analyses into the System-level analysis.

There are multiple benefits to the BXM method:

1. It allows decomposing a complex System into several simpler parts which are easier to understand, and less prone to errors of omission or commission during the analysis.
2. It allows parallel work streams where different teams can be simultaneously analyzing the different System components.
3. It enables the management to assign people with relevant skills to the appropriate teams. For example, during analysis of a *mechanical* component of the System, *electrical* engineers would not be required to attend, as they would not add significant value to that analysis.
4. Modularity and reuse. If a particular component is used in multiple Systems, or even different generations of the same System, the analysis of that component can be reused.

In the following sections the BXM method is elaborated.

10.1 SYSTEM DECOMPOSITION

Let's say that at the top-level the System is called Level-1 (L1). In the example in Fig. 10.1 the L1 System is decomposed into two Level-2 (L2) components. Each L2 component is further decomposed into multiple L3 components and so on. This decomposition follows the System architecture.

Safety Risk Management for Medical Devices
DOI: https://doi.org/10.1016/B978-0-12-813098-8.00010-6

© 2018 Elsevier Ltd.
All rights reserved.

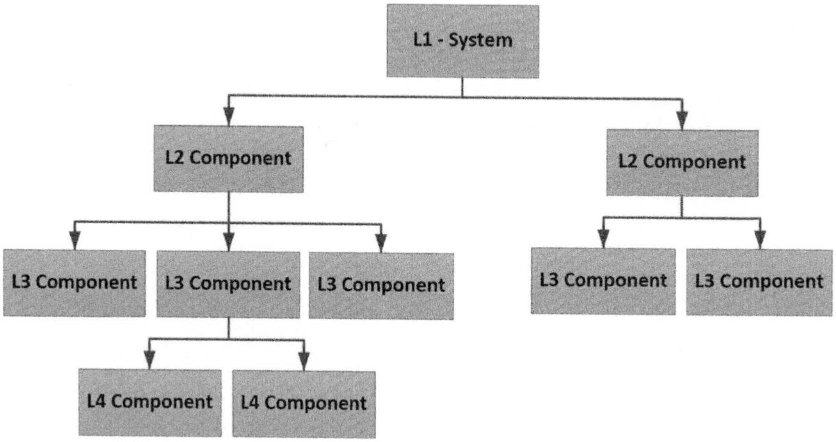

Figure 10.1 System decomposition.

The criteria for decomposition and how far to go, are: the novelty of the System, and the degree of reusability that you want. For novel Systems, decompose to a lower level. If a component is reused in other Systems, then decompose to a level where the reusable component gets analyzed. That way, its analysis would become available for reuse elsewhere. To elucidate, two examples are offered.

Example 1—We are analyzing an automobile for safety. The fuel system in this automobile has been in use in multiple models and there is a lot of performance data available on it. As we decompose the automobile, when it comes to the fuel system we do not further decompose it, because of the knowledge and history that is available on it.

Example 2—Let's say an automobile manufacturer uses the same brake caliper in a brake system which is in use in three different automobile models. The brake system is well understood and has a history of use in the field. So, ordinarily we would not need to decompose the brake system further. But, we are going to design a new brake system which will use the same caliper. We don't already have an analysis for the caliper. We want to reuse the analysis of the brake caliper in the future brake system. So, in this case we would decompose the brake system down to the level of the caliper.

10.2 INTEGRATION

Integration is the corollary and complementary concept to decomposition, which was described in Section 10.1. The principal concept in integration strategy is the hierarchical multilevel Failure Modes and Effects Analysis (FMEA). See Sections 12.4.2 and 16.1 for details of this mechanism.

The BXM methodology uses the architectural design of the System as a road map, and looks for the Failure Modes of each architectural element. A critical principle in this method is the strict adherence to the scope, and boundary of analysis within the FMEAs. This principle allows the performance of an FMEA of a given component agnostic of the System in which it is used. That is, the FMEA needs to only concern itself with identifying the End Effects at the boundary of analysis. A great benefit of this principle is that the analyses of the architectural components can be integrated in the same way that the physical components are integrated per the System architecture.

10.3 QUANTITATIVE RISK ESTIMATION

Another attribute of the BXM method is the quantitative estimation of risk. See Section 17.3 for details. Quantitative estimation of risk enables a simple way of evaluating the acceptability of residual risk. It boils down to a simple comparison of two numbers: the residual risk, and the acceptable risk level. The BXM method uses Boolean algebra to compute the residual risk of a System: per-Hazard, per-Hazardous-Situation, and overall.

Thanks to its mathematical approach, the BXM method lends itself to implementation in software tools. The benefits of use of a software tool in risk management are:

1. Objective and automatic determination of risk acceptability
2. Avoidance of error-prone manual computation/assessment of risk
3. Ability to always have an up-to-date risk assessment
4. Ability to evaluate the safety impact of proposed design changes
5. Ability to reuse estimations of Harm probabilities across multiple projects
6. Ability to compute the overall residual risks of the System (medical device)

CHAPTER 11

Risk Management Process

Abstract

ISO 14971 requires that the manufacturer have a documented process for risk management, and provide the specifications for such a process. However, the Standard does not provide a process—that is left up to the manufacturer. In this chapter a risk management process is presented, and expounded. One important aspect of any risk management process is the criteria for risk acceptability, which must be identified in the Risk Management Plan.

Keywords: Risk Management Process; management responsibilities; risk management file; risk management plan; criteria for risk acceptability; clinical hazards list; harms assessment list

ISO 14971 [3,7] requires that the manufacturer have a documented process for risk management, and provide the specifications for such a process. However, the Standard does not provide a process—that is left up to the manufacturer.

If risk analyses for a similar medical device are available, relevant, and adequate, they can and should be applied to the analysis of the subject medical device.

The BXM method uses a process that is depicted in Fig. 11.1. In broad terms, it includes Hazard identification, risk estimation, risk control, risk evaluation, and monitoring.

This process is applicable for:

- each new device, or derivative device
- new indications for an existing device
- each change (in a part) of a released device
- each change (in a part) of a realization process of a released device, including changes to suppliers/manufacturing sites
- discovery of mislabeled or nonconforming product
- Corrective and Preventive Actions (CAPA) events with potential risk to patient safety

The process starts with formation of the Risk Management File (RMF; see Section 11.2), and writing a Risk Management Plan (RMP; see Section 11.3). It would be beneficial to determine the safe-state(s) of your medical device, if any. This would help both in the design and in the risk management decisions. Next, a Preliminary Hazard Analysis (PHA) is done (see Section 12.2). Thereafter, the project transitions to the design and development phase.

As designs become available Failure Modes and Effects Analyses (FMEAs) are performed and iterated throughout the design and development phase (see Section 12.4).

DOI: https://doi.org/10.1016/B978-0-12-813098-8.00011-8

© 2018 Elsevier Ltd.
All rights reserved.

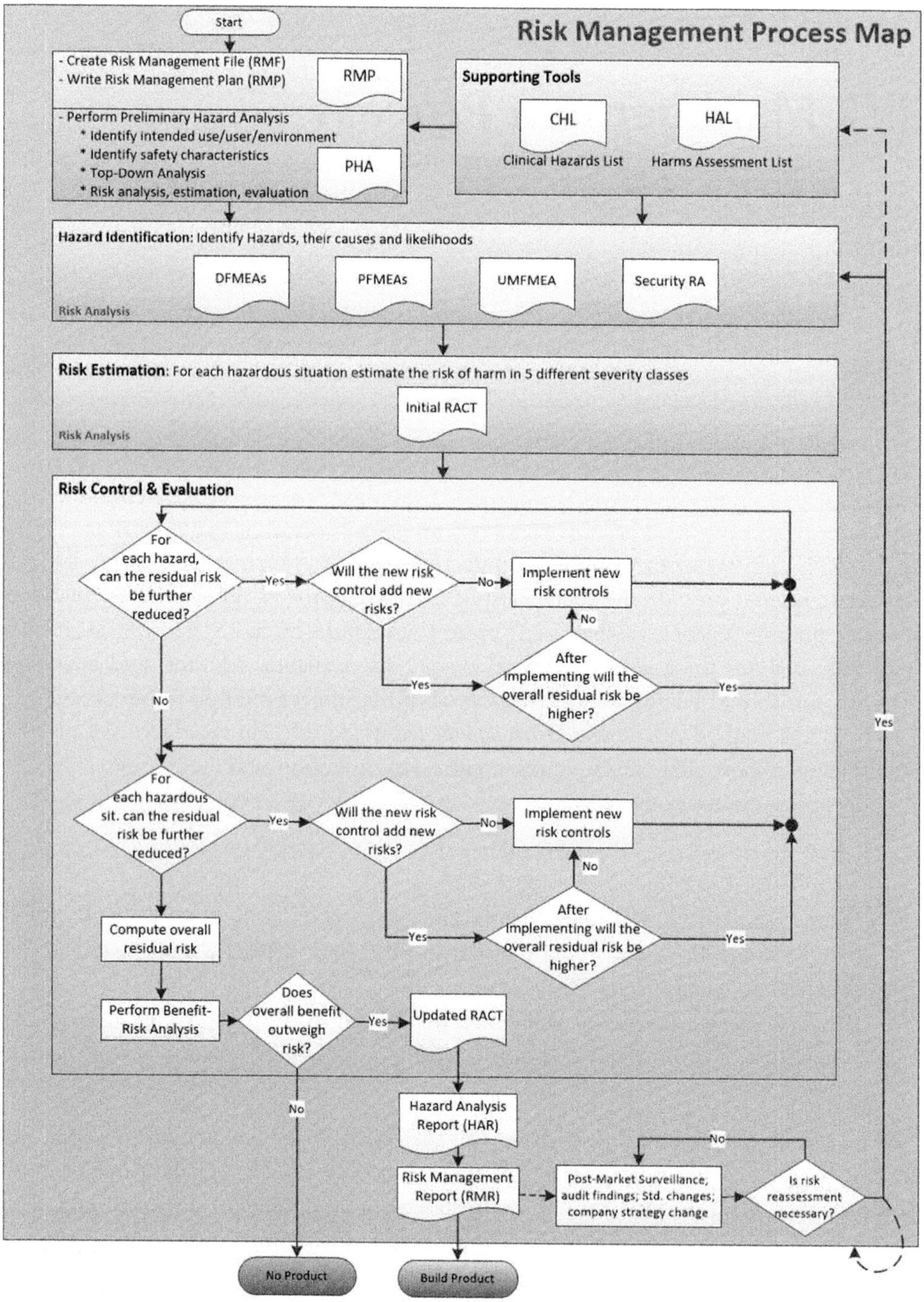

Figure 11.1 The BXM risk management process.

This together with Fault Tree Analysis constitutes the Hazard identification phase. Additionally, the Security Risk Assessment file is examined for potential security risks with a safety impact.

Next, the risk estimation phase is entered (see Chapter 17: Risk Estimation). This is where all the Hazards and their causes, the corresponding Hazardous Situations and Harms are brought together in one table called the Risk Assessment and Control Table (RACT). Risk Controls should be applied, and risks reduced as far as possible (in compliance with EN ISO 14971:2012 [7]). Within the RACT, the risks for each Hazard, Hazardous Situation as well as overall, for the whole System are estimated. The RACT is the heart of the BXM risk management process. In any ISO 14971 compliant risk-management process something like the RACT is found. It may be called by other names, e.g., Risk Table, Risk Matrix, Risk Analysis Chart, etc.

Following the Risk Controls and risk estimation, another pass is made to investigate whether it is possible to further reduce the residual risks for each Hazard or Hazardous Situation. If possible, additional Risk Controls are implemented. Next, the overall residual risk for the System is assessed with consideration of *all the Hazards* that the System could present.

Once the overall residual risk of the System is known, a benefit–risk analysis is performed to determine whether the potential benefits of the System outweigh its potential risks. If the benefits do not outweigh the risks, and no justification can be made for release of the product, then the System should not be released for commercial purposes. If the benefits do outweigh the risks then update the RACT with any additional Risk Controls that were put in place since the initial RACT was produced, and produce a Hazard Analysis Report (HAR). A HAR is a relatively large document which embodies much of the details of the risk management process, e.g., the RACT and the benefit–risk analyses.

For submission purposes, an executive summary of the HAR is produced as input to the Risk Management Report (RMR). While the HAR could be overwhelming to a Regulatory reviewer due to its size and depth of detail, the RMR is a smaller document that is designed to provide a good understanding of the risk management work and give the Regulatory reviewer confidence that the System is safe enough for commercial release.

The RMR is included in the Regulatory Submission. After the approval and release of the medical device, the risk management process continues for as long of the medical device is in the market. Periodically, the risk management work-products are examined for potential impact due to input from production and postproduction monitoring. Input such as complaints, trending, other field data, and even changes to the Standard ISO 14971 [3,7] itself are considered. If any change or finding warrants revising of the risk management work-products, the required changes are done and the results are reflected in the RMR.

ISO 14971 [3] does not specify the periodicity of production and postproduction data review. The manufacturer gets to specify that in the RMP. But the choice by the manufacturer must be defensible. A reasonable approach would be to choose the period based on the device risk, and novelty. For example, if a device is high risk and new, it makes sense to review its RMF more frequently immediately after launch, e.g., twice a year. Then as more knowledge is gained about the device, reduce the frequency to once a year or once every 2 years. For devices that are just iterations on an existing device, about which much knowledge exists, you can start with a lower frequency. In any case, if an adverse event happens in the field, or changes are made to the design or indications of the device, the RMF must be examined for potential impact.

11.1 MANAGEMENT RESPONSIBILITIES

ISO 14971 [3,7] defines specific responsibilities for Top Management. "Top Management" is not defined in the Standard [3,7], and could mean different things in different companies. For example, in a small company Top Management could be the CEO. But that would not be the case in a large multinational company. The best way to discern what "Top Management" means for your company is to consider the governing Quality Management System (QMS). The person(s) at the level where Risk and Quality policy decisions are made, can be considered "Top Management." Therefore for a company with multiple business units that have different QMSs, "Top Management" could be the business unit board members, e.g., manager of R&D, Manager of Quality, and the general manager.

Top Management is required to provide evidence of its commitment to the risk management process by:

- ensuring the provision of adequate resources
- ensuring the assignment of qualified personnel for risk management

Other responsibilities of Top Management are:

- define and document the policy for determining criteria for risk acceptability (see Section 11.3.1 for further details)
- review the suitability of the risk management process at planned intervals to ensure continuing effectiveness of the risk management process. Any decisions and actions taken during the reviews should be documented. If you have a QMS in place, e.g., one that is compliant with EN ISO 13485 [21], the risk management process review can be part of the regular management reviews of the QMS.

11.2 RISK MANAGEMENT FILE

One of the earliest actions in the risk management process is the creation of a RMF. The RMF is a repository of the risk management artifacts. The purpose of the RMF is to enable easy and prompt access to the risk management artifacts. The RMF can take any form that supports its purpose. For example, the RMF could be in paper form, in folders and cabinets, or it can be in electronic form as a collection of files in a computer. It can even be an index to files that are in different locations.

The RMF should have a keeper—a person who is responsible for its upkeep and maintenance. This is particularly important after product development is concluded and the product is released. This is because risk management is a living process which continues for as long as the product is in the market. Therefore the risk management artifacts are periodically updated and need to be properly filed in the RMF.

The Standard [3,7] does not prescribe a list of items that must be in the RMF. But the following is a potential list of items that could be found in an RMF:

- RMP
- Top-Down analyses, e.g., Fault Tree Analysis
- PHA
- Design FMEAs
- Software FMEAs
- Process FMEAs
- Use-Misuse FMEA
- RACT
- RMR(s)
- Records of reviews and approvals of risk management artifacts
- Clinical Hazards List (CHL)
- Harms Assessment List (HAL)
- Benefit–risk analyses
- Verifications of implementation of Risk Controls
- Verifications of effectiveness of Risk Controls
- Log of production and postproduction activities

11.3 RISK MANAGEMENT PLAN

ISO 14971 [3,7] requires that a RMP be created at the beginning of the risk management process. The RMP should include the following elements:

1. Purpose and scope
   - Scope should identify what phases of product development process are addressed in the plan. It is possible to initially write a plan for the early

phases of the product development process, and later update the RMP to include the remaining phases.

2. Overview of the System
   - Describe the System, its function, its elements, indications, intended use, user, and use environment.
   - Identify what is included in the analysis. It's a common mistake to exceed the scope of analysis and include peripheral devices that are not part of the System in the analysis.
3. State your risk management strategy
   - What's your main strategy to make your System as safe as possible? Examples:
     - For a therapy-advisory System, you may choose to keep the physician in the decision-making loop.
     - For a deep brain stimulator, you may require the use of accurate navigation in the brain to avoid causing brain hemorrhage during the implant surgery
4. List your planned risk management activities
   - Examples: PHA, Fault Tree Analysis, FMEA, Benefit–risk analysis, etc.
5. Identify any special tools, such as FTA software, customized software, etc.
6. Identify people/roles who have responsibility for risk management activities, and their authorities. Include who will be responsible for the maintenance of the RMF.
7. Spell out the requirements for review of risk management activities.
8. Define the risk acceptance criteria.
9. Describe the verification activities and deliverables. This includes both verification of implementation of the Risk Controls, and verification of effectiveness of the Risk Controls.
10. Describe how risk management will affect other aspects of product development process such as sample size determination, production acceptance criteria, etc.
11. Describe how production and postproduction information will be captured and fed back into the risk management process.

The above list makes for a good RMP. However, the minimum requirements for the RMP per ISO 14971 [3,7] are items 1, 2, 6, 7, 8, 9, and 11.

11.3.1 Criteria for risk acceptability

ISO 14971 [3,7] clause 3.2 requires top management to define and document the policy for establishing and reviewing the criteria for risk acceptability. The policy should ensure that the criteria:

a) are based upon applicable national or regional regulations
b) are based upon relevant international standards
c) take into account available information such as the generally accepted state-of-the-art (SOTA) and known stakeholder concerns

The policy for determining the criteria for risk acceptability could apply to the entire range of a manufacturer's medical devices, or be specialized for different groupings of the manufacturer's medical devices. The manufacturer can take into account the applicable regulatory requirements in the regions where a medical device is to be marketed.

The relevant International Standards for a particular medical device, or an intended use of a medical device can help identify the principles for setting the criteria for risk acceptability.

Information on the SOTA can be obtained from review of published scientific literature. Also, information on similar medical devices that the manufacturer has marketed, as well as medical devices from competing companies can serve to identify the SOTA.

With respect to stakeholder concerns, many sources can be utilized, including: news media, social media, patient forums, key opinion leaders, as well as input from internal departments such as regulatory, clinical, and marketing who have knowledge of stakeholder concerns.

Criteria for risk acceptability provide guidance for determination of end-points for risk reduction. ISO 14971:2007 [3] requires the reduction of residual risk to As Low As Reasonably Practicable, while EN ISO 14971:2012 [7] requires reduction of residual risk to As Far As Possible, which also sometimes referred to as As Low As Possible, without regard to economic considerations.

The NBRG Consensus Paper [22] disputes the assertion of Ref. [7] on no allowance for economic consideration. The NBRG consensus paper [22] states that "Although economic considerations will always be relevant in decision-making processes, the safety of the product must not be traded off against business perspectives. For transparency, the manufacturer must document the end-point criteria of risk reduction based on his risk policy."

The requirement of Ref. [7] that risks be reduced to as low as possible for acceptability creates both a nebulous and an impractical criterion for risk acceptability because first, it is difficult to know when the lowest possible risk level has been reached (short of total elimination), and second, trying to achieve this mark could restrict patient access to safe and affordable medical devices. In response to Ref. [7], Ref. [22] provides a practical interpretation of the Standard [7], as follows:

If "death or serious deterioration of health is unlikely to occur in normal operation or due to device malfunctions or deterioration of characteristics or performance, or any inadequacy in the labeling or instructions for use", then "the risk shall be considered acceptable" [22].

Otherwise the risk must be reduced. Ref. [22] offers some options to determine end-points for risk reduction:

1. Conformance to a harmonized product safety standard. When such a standard specifies technical requirements addressing particular Hazards or Hazardous Situations, together with specific acceptance criteria, compliance with those requirements can establish a presumption of acceptable risk.
2. In the absence of an applicable harmonized standard, basing the risk reduction end-point on a recognized international standard would also constitute acceptable risk. Depending on where the medical device is going to be marketed, it is also possible to apply a similar strategy which is specific to a geography. That is, it's possible that a particular country has a national product safety standard. Therefore compliance to that standard would establish acceptability of risk in that country.
3. Where no harmonized standard, international or national standard, or similar publication is available, the manufacturer can use other sources such as historical data, best medical practice, and SOTA as means of specifying risk reduction end-points. Methods of determination of end-points must be described in the risk management process.
4. Lastly, risk reduction end-point can be defined as when further Risk Control measures do not reduce risk, or improve safety.

It must be noted that if advancement in technology and practices that exist, and are feasible at the time of design of a medical device, permit the reduction of risk beyond the SOTA, it is incumbent upon the manufacturer to apply such new knowledge and methods and further reduce the risks of the medical device.

The RMP for each medical device must provide the specific criteria for risk acceptability consistent with the manufacturer's policy for determining the criteria for risk acceptability.

Although Ref. [22] is a tremendous improvement over Ref. [7] and greatly reduces the ambiguity of Ref. [7], it still leaves some unclarity. For example, Ref. [22] says if "death or serious deterioration of health is unlikely to occur..." then "the risk shall be considered acceptable." What does "unlikely" mean? There is no guidance on how to decide. Also, the guidance from Ref. [22] doesn't prompt the manufacturer to reduce risks to lower levels than the SOTA, when new technologies feasibly allow doing so.

The BXM method offers a risk reduction end-point logic that is consistent with, and an improvement over Ref. [22]. It gives more clarity on how to decide on risk reduction end-points where there is a SOTA, and also where there is no SOTA.

Fig. 11.2, is applicable when there is a SOTA. The logic should be applied to all Hazards and repeated until no further risk reduction is called for.

Fig. 11.3 shows a condition where there is no SOTA. The logic should be applied to all Hazards and repeated until no further risk reduction is called for.

For some medical devices of small significance, there may not be any published scientific literature reporting clinical studies of those devices and the corresponding safety outcomes. In such cases, it may not be feasible to determine SOTA. An

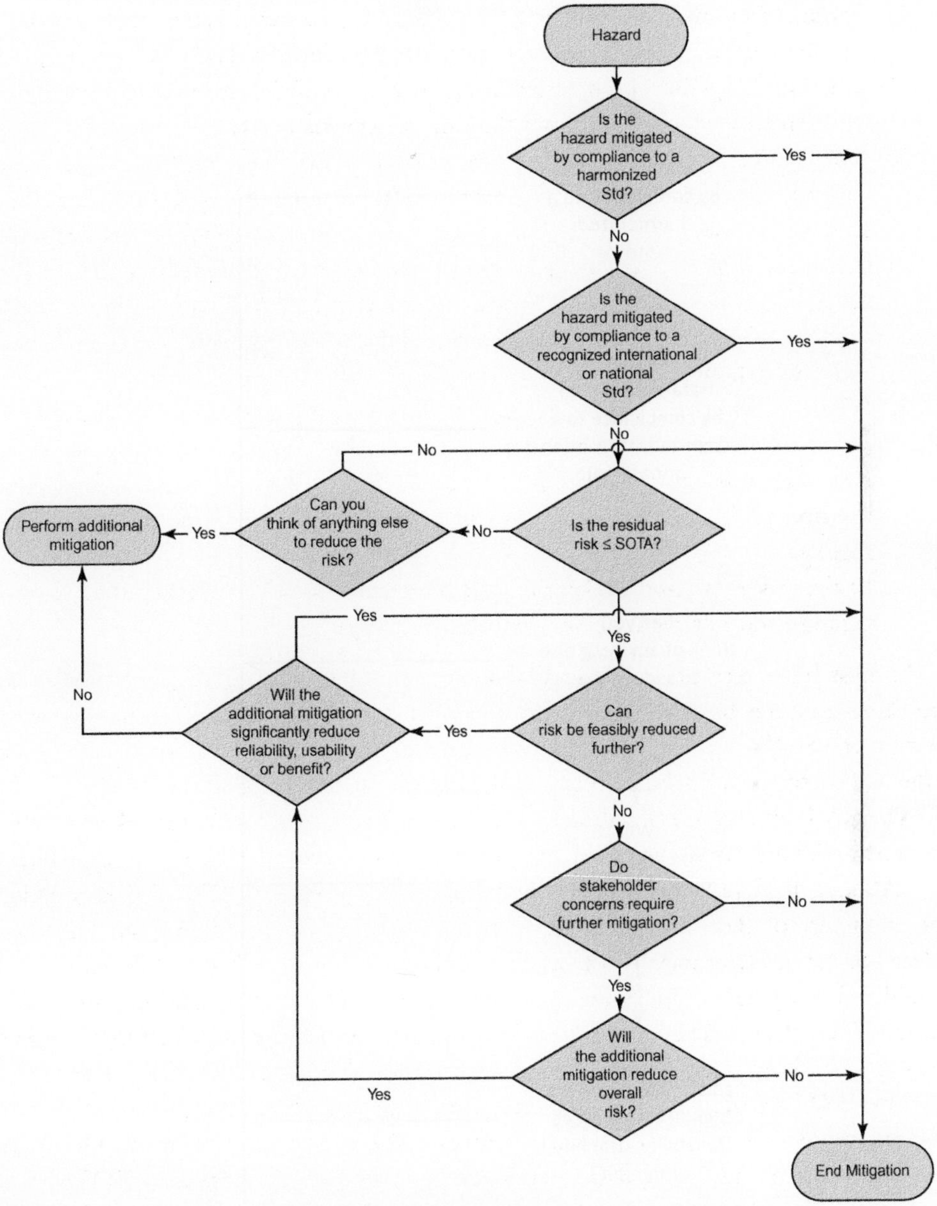

Figure 11.2 Risk reduction end-point logic (with SOTA).

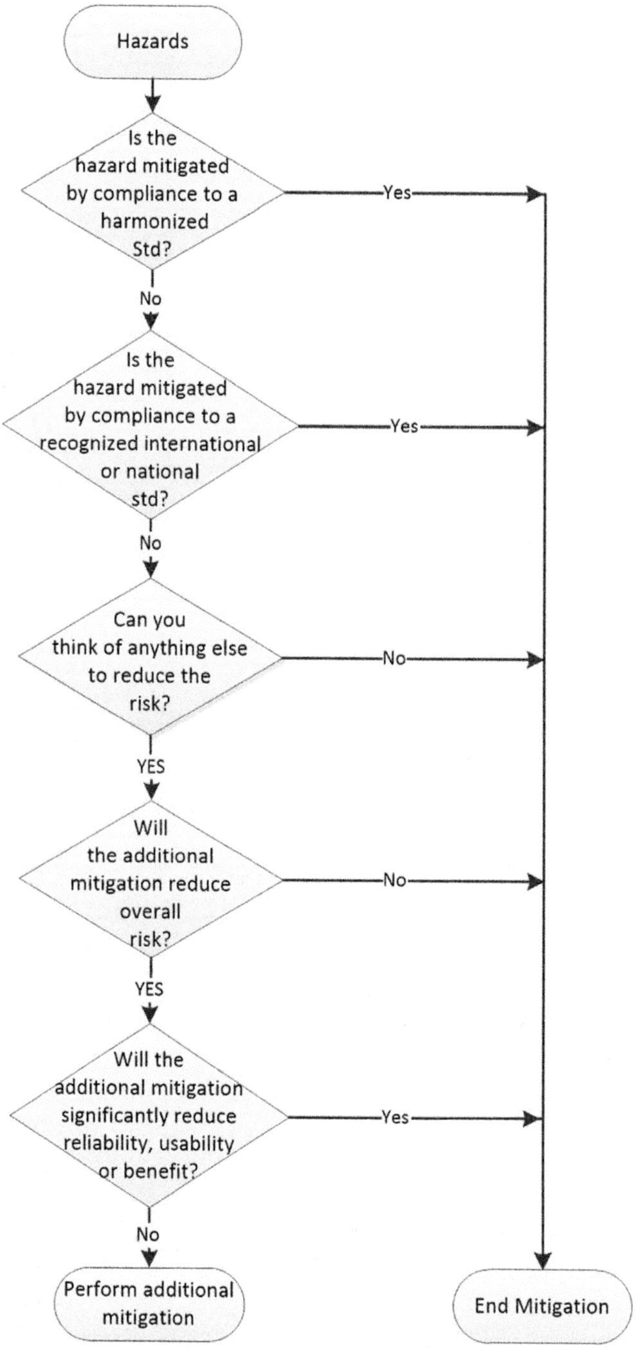

Figure 11.3 Risk reduction end-point logic (without SOTA).

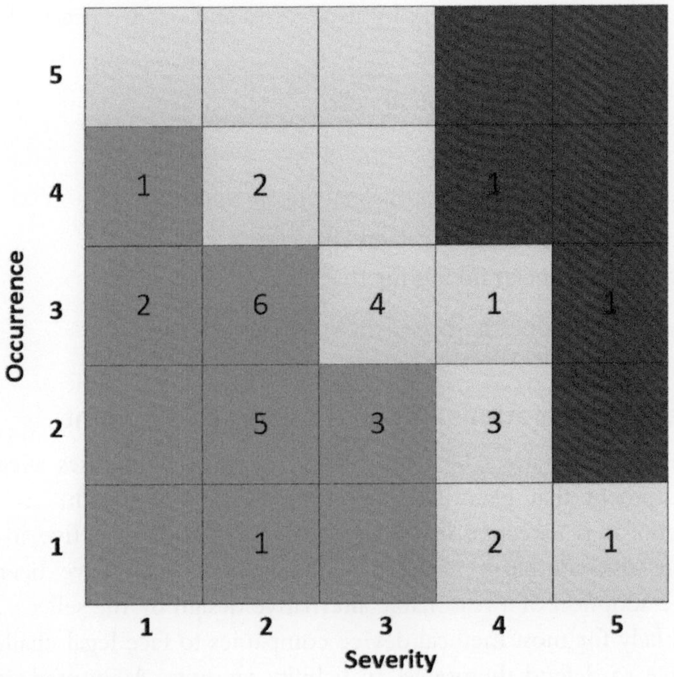

Figure 11.4 Example risk profile.

alternative method for such cases is to create a safety risk profile for the device and follow the normal benefit–risk analysis process. If a panel of medical experts consider the benefits of the device outweigh the overall residual risks of the device, and the device gets approved for commercialization, then the risk profile of that device can serve as the basis for evaluating future versions of the same product.

An example risk profile could look like Fig. 11.4. The method to create this type of risk profile is as follows.

Use a severity scale such as Table 17.1, and a probability scale such as Table 17.2. Determine the risk of Harm(s) for every Hazard. If your process uses a single Harm-severity method, then *R* describes the probability of occurrence of the Harm in *one* class of Harm, e.g., Critical (rank 4). For all the Hazards of the System, count the number of risks that fall in each cell of the matrix in Fig. 11.4. Populate the matrix as in Fig. 11.4. This would be the risk profile for the device in question and could serve as the basis of risk evaluation for the next version of the same device that is released.

In the BXM method, where for each Harm there are five P_2 numbers, one for each severity class, the same computation can be implemented. Only instead of one *R* for each Hazard–Harm pair, there would be *five* Rs. The counting method is the same as above.

To compare one risk profile against another, you need an algorithm. One potential algorithm is to assign a score to the risk profile which is the sum of the products of each entry in each cell of the matrix by its severity and occurrence rankings. In other words:

$$\text{Score} = \sum (N \times S \times O) \text{ for all cells:}$$

where,

N = the entry in the cell
S = the severity ranking for that cell
O = the occurrence ranking for that cell

A lower score would be better than a higher score.

11.3.2 Other considerations for risk reduction end-point

Bordwin in Ref. [23] states "The law books are filled with cases alleging defective design of equipment that plaintiffs claim caused injury or death. "; "A product is defective when...it is defective in design.... A product is defective in design when the foreseeable risks of harm posed by the product could have been reduced or avoided by the adoption of a reasonable alternative design by the seller..."

It is very likely for most medical device companies to face legal challenges at some point, and have to defend themselves in liability lawsuits. As quoted earlier, in such lawsuits the assertion would be that the manufacturer did not take reasonable measures to reduce the risk of Harm from the device. Having a rational, defensible, and documented methodology to determine risk reduction end-point, would be very valuable in the legal defense in such lawsuits. In particular, as Ref. [22] states "safety of the product must not be traded off against business perspectives."

11.4 HAZARD IDENTIFICATION

The manufacturer is expected to identify all risks due to the use of the medical device. This requires knowledge of all the *Hazards* associated with the use of the device. While it is difficult to claim completeness, the use of a standardized CHL enables you to claim that you are as complete as possible. See Section 11.5 for further details on the CHL and how to create it.

The Hazard identification phase also entails identifying the Causes of the Hazards and the likelihood of occurrence of the Hazards. This is facilitated by the use of the technique of FMEA. See Section 12.4 for details on FMEAs.

A major source of Hazards is the interoperability of interfacing parts. Most medical devices are comprised of part that connect and interoperate with each other. These include mechanical, electrical, and informational interfaces. Additionally, you should consider interfaces with the outside world.

An important, and sometimes missed aspect of medical device risk analysis is the packaging. Packaging must protect the medical device in handling, transportation, and storage. For sterile products, the packaging also has the duty of maintaining sterility of the product. Packaging also has the role of protecting people, property, and the environment from the device.

In addition to the Hazards that could directly harm patients, also consider Hazards that could indirectly harm patients. An example of indirect Hazard is the informational type. For instance, if a diagnostic device produces a false negative result, it could mislead the clinician into not delivering the needed therapy to the patient.

11.5 CLINICAL HAZARDS LIST

A CHL is a complete list of *all* the known and foreseeable Hazards that could potentially arise from the use of the medical device. A claim of completeness is special and noteworthy. How this claim can be made is by the mechanism with which the CHL is created and kept up to date.

A good place to start for identifying the Hazards associated with the medical device is Table E.1 of ISO 14971 [3,7]. Use this table as a thought stimulator. Likely you will find many of the Hazards are irrelevant to your device, but on the other hand it might bring to your attention, some Hazards that you could have overlooked. Table B.2 of IEC 62366 [14] is another resource that could serve as a thought stimulator for identification of Hazards due to use errors. Next, examine other sources. Some suggestions are as follows:

- Literature search of comparable products
- Your internal historical data based on CAPAs and complaint handling database
- Adverse-events databases, e.g., MAUDE [16], EUDAMED [17]
- Input from subject matter experts

After you have compiled the list of Hazards, review it for duplications, and erroneous entries. It is common that people enter items in the CHL that are not Hazards. A common mistake is to enter Causes, Hazardous Situations, or Harms in the list. Use this test to distinguish true Hazards: "If someone was exposed to this item, could they potentially be harmed?"

Examples of CHL entries that are not Hazards:

- Missing labeling
- Software
- Bleeding

Table 11.1 Example of Harms Assessment List

ID	Harm	FDA Code	Severity Class: Catastrophic (%)	Critical (%)	Serious (%)	Minor (%)	Negligible (%)	Total (%)
Harm.1	Burn (thermal tissue heating)	1757	1.5	4.0	66.8	19.8	7.9	100
Harm.2	Cerebral Fluid Leak	1772	0.1	2.3	60.4	35.8	1.4	100
Harm.3	Hemorrhage, intracranial	1891	2.5	19.0	31.7	46.8	0.0	100
Harm.4	Infection	1930	0.9	2.2	72.1	18.3	6.5	100

After the first pass of cleanup, route the list to subject matter experts. For example, clinicians, clinical investigators, or R&D engineers who have experience with similar devices. After the second round of reviews, the list should be ready for approval and use in your risk management process.

You may find that some Hazards may need to be stratified. For example, cessation of breathing for 10 seconds, versus 2 minutes, versus 5 minutes could have dramatically different harms. In such cases the same Hazard could be cited n times, each with a different qualification.

The CHL is a living document which is kept up to date. If at any time a new Hazard manifests that was not previously in the CHL, revise and update the CHL. This is how the claim of completeness can be made—that at any given time, the CHL is as complete as possible.

11.6 HARMS ASSESSMENT LIST

Another pivotal tool in the BXM method of risk management is the HAL. The HAL is a list of all the potential Harms that could result from the use of the System under analysis. Table 11.1 shows an example of a partial HAL.

Note: The data in Table 11.1 is fictitious—do not use for actual analysis.

The HAL provides P_2, the probability of sustaining Harm in a Hazardous Situation. In other words, it is assumed that the Hazardous Situation has already happened.

The HAL model presented here is inspired by ISO 14971 [3,7]. Note the following in the construct of the HAL:

- Harms are physical injury.
- For each Harm five P_2 numbers are provided, one for each Severity class.
- Sometimes no Harm occurs in a Hazardous Situation. In this example, the lowest Severity class: Negligible includes the "No Harm" outcomes.

- The totals of all the five P_2 numbers add up to 100%. This means all possibilities are accounted—from nothing to death.
- The Harm outcomes are the aggregate of all potential circumstances, e.g., for Harm.4 in Table 11.1, if some infections receive medical care and some don't, the cited probability of Harm numbers account for both possibilities.

The creation of a HAL requires foresight and good judgment. Harm outcomes depend on many factors. The same Harm would have different outcomes on different patient populations, under different conditions, etc. For example, consider the Harm of electrocution, meaning injuries due to electric shock. The voltage and current of the source, the impedance of the skin of the person, the location of electrical discharge on the body are all factors on the severity of the Harm. The creator of the HAL may choose to create multiple lines for the Harm of electrocution, e.g.,

Electrocution due to:

- Exposure to 100–200 V
- Exposure to 200–400 V
- Exposure to 400^+ V
- Exposure to unprotected hand
- Exposure to gloved hand
- Exposure while user is standing in water
- etc.

Application of this strategy to all the Harms could create a very large HAL. Alternatively, all the various circumstances of electrocution could be aggregated into one line—electrocution. This option would offer less precision in the determination of the risk of Harm, but is more practical with respect to the size of the HAL. A midway method would be to consult KOLs about the most likely levels of Harm, and stratify the Harm only to those most likely levels. Striking the right balance is the art of risk management.

Remember that only a qualified person (e.g., an MD) can assign severities to Harms.

11.6.1 How to create a Harms Assessment List

There are two methods by which a HAL can be created:

Method 1—Using published data
Method 2—Using expert opinion

Each method is detailed as follows.

11.6.1.1 Method 1—Using published data

Method 1 is the preferred method because it relies on actual data. The steps to follow for Method 1 are as follows:

1. Define search criteria, and exclusion criteria for the published papers.
2. Search for published papers, or other formal sources using the search criteria from Step 1. There are many databases such as PubMed, or Cochrane that can be used for this purpose.
3. Filter the results of the search based on the exclusion criteria.
4. Review the remaining papers to ensure appropriateness of the selected papers for your System.
5. From each paper extract the number of observations of the Harms in the HAL in each severity class of: (Catastrophic, Critical, Serious, Minor, Negligible)
6. Sum the total number of observations per severity class, per harm from all papers. This makes the numerator for P_2 computations.
7. For each Harm in the HAL sum all the observations in all severity classes from all the selected papers. This makes the denominator for P_2 computations.

Example: Let's say we have selected 14 published papers that are relevant to the System under analysis. And, let's say the Harm of interest is Hemorrhage, intracranial.

The total number of reported intracranial hemorrhages in all papers = 79
Total number of deaths from intracranial hemorrhage (catastrophic) = 2
Total number of symptomatic, persistent intracranial hemorrhages (critical) = 15
Total number of symptomatic, treated transient intracranial hemorrhages (serious) = 25
Total number of asymptomatic untreated intracranial hemorrhages (minor) = 37

In no case the physicians considered an intracranial hemorrhage an inconvenience/discomfort (negligible).

Therefore,	P_2 (catastrophic)	= 2/79	= 2.5%
	P_2(critical)	= 15/79	= 19.0%
	P_2(serious)	= 25/79	= 31.7%
	P_2(minor)	= 37/79	= 46.8%
	P_2(negligible)	= 0/79	= 0.0%
	Total		= 100%

Tip If you have access to registry data, it's a preferred source over published clinical studies. The reason is that in clinical studies in many cases patients are selected who are most likely to benefit from the therapy. Whereas in registries all users of a device are taken into consideration. And, the patient population is much larger than those in clinical studies. Therefore the data is more representative of the "real world." Also, in comparison with complaint data, registry data is superior because for registries data is actively sought, while complaint data-gathering is passive and likely subject to underreporting.

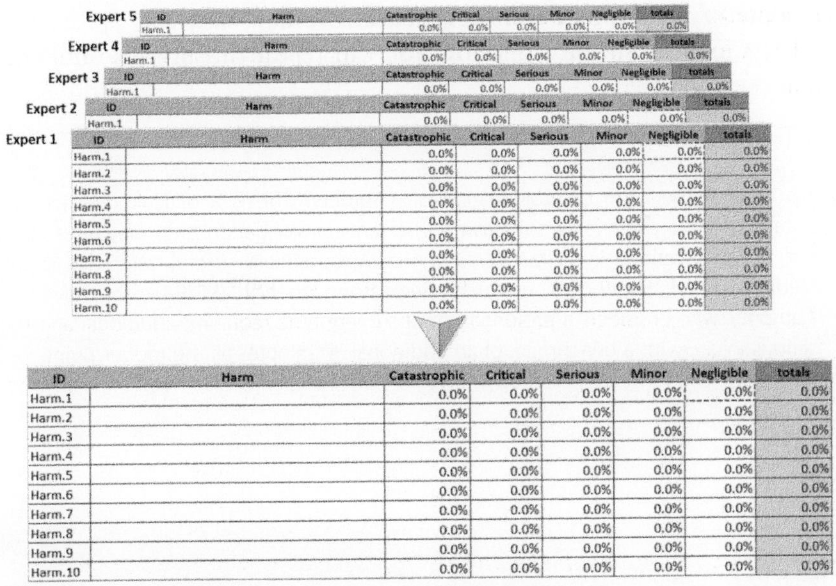

Figure 11.5 Harms Assessment List creation via expert opinion.

11.6.1.2 Method 2—Using expert opinion

In this method, you seek the opinions of multiple experts and ask them the question: *given that the Harm has happened*, what is the likelihood of:

- Death
- Permanent impairment or life-threatening injury
- Injury or impairment that requires professional medical intervention
- Temporary injury or impairment that does not require professional medical intervention
- Inconvenience, or temporary discomfort or no harm

Ask this question for every Harm that is listed in the HAL. Then aggregate the responses of all the experts into an overall table. Fig. 11.5 displays a graphical representation of this method.

Method 2 is inferior to Method 1, because it depends on the subjective opinion of clinicians. The opinion of a clinician is formed by their personal experience with their patients, which is naturally limited. The more clinicians that are interviewed, the better the quality of the data in the HAL.

Delphi technique—As an extra means of improving the quality of the HAL, it is recommended to show the aggregated HAL to the interviewed clinicians, in a second round of interviews. It may be that they would change their minds on their initial estimates, once they see the aggregate of the collective opinions. If they change their mind on some of their initial estimates, collect the new data and update the HAL with the latest input from the clinicians. You can repeat this process until consensus is achieved.

Tip What is the harm of prolongation of a surgery? This is a difficult question to answer because no two surgeries are the same. The same prolongation time may have vastly different Harm severities in different surgeries. A potentially useful reference is from the Medicines and Healthcare Products Regulatory Agency (MHRA)—the UK Competent Authority, who considers a prolongation of surgery time requiring additional anesthetic as a serious injury, with a benchmark of an additional 30 minutes as the trigger point.

CHAPTER 12

Risk Analysis Techniques

Abstract

Identification of hazards for risk analysis can be done using various tools. Two of the most common tools are Fault Tree Analysis (FTA) and Failure Modes and Effects Analysis (FMEA). In this chapter three types of FMEA are discussed: DFMEA, PFMEA, and UMFMEA. Additionally, two other tools are presented: Mind Map Analysis and P-Diagram. Ultimately it is the analyst's choice on how many tools to use. While extra analyses consume more resources, they also reduce the likelihood of missing some Hazards and their causal chains.

Keywords: Fault Tree Analysis; FTA; Failure Modes and Effects Analysis; FMEA; DFMEA; PFMEA; UMFMEA; Mind Map Analysis; P-Diagram

12.1 FAULT TREE ANALYSIS

12.1.1 Introduction

The Fault Tree Analysis (FTA) technique was developed by Bell Labs in 1962 for use on the Minuteman missile system. Later it gained wide use in civil aviation, space, and military applications. MIL-HDBK-338B published in 1998 provides a reference for this technique. After the 1979 incident at Three Mile Island, the Nuclear Regulatory Commission expanded the use of FTA and published NUREG-0492—*Fault Tree Handbook* in 1981. This handbook was later updated by NASA in 2002 with the title *Fault Tree Handbook with Aerospace Applications* [24].

FTA is a deductive top-down reasoning process that starts from the undesired system outcomes and attempts to find out all the credible sequences of events that could result in the undesired system outcomes. The fault tree is a graphical model that depicts the logical relationships among the parallel and sequential combination of events that could lead into the event at the top of the tree.

FTA can model both normal and fault conditions, under various environmental and operational scenarios. FTA can also identify and model fault dependencies, and common cause failures (CCFs).

Fault Trees (FTs) are constructed using logic gates, such as AND and OR gates. As such, FTs lend themselves to logical simplification and reduction. Therefore there is not just one correct FT to describe a system, but potentially multiple logically equivalent FTs.

Safety Risk Management for Medical Devices
DOI: https://doi.org/10.1016/B978-0-12-813098-8.00012-X

© 2018 Elsevier Ltd.
All rights reserved.

Due to its nature, the FTA can be utilized for quantitative analysis, to estimate the probability of occurrence of the top undesired events. It is important to remember that an FT is not in itself a quantitative model. It is a qualitative model that can be evaluated quantitatively.

The FTA can be applied to new products before design details are available. In this capacity, the FTA can reveal at a high level, potential event sequences that could result in System Hazards, and thus alert the design teams to safety-critical aspects of the System. When FTA is applied to existing Systems, it can identify design weaknesses and aid in the identification of design upgrades to make the System safer.

A principal output of the FTA is a collection of minimal cut sets to the top event. "A cut set is a set of basic events, which if they all occur, will result in the top event of the fault tree occurring" [24]. A minimal cut set is the smallest set of basic events, which if they all occur will result in the occurrence of the top event. The term minimal is used to mean that in a given path to the top event, if any of the basic events doesn't happen, then the top event won't happen. Minimal cut sets can also be identified for intermediate events. If a minimal cut set is shown to be comprised of only one basic event, it reveals a vulnerability to a single-point failure.

When probabilities are applied to a FT, the dominant cut sets can be identified. Those are the pathways with the highest probabilistic contribution to the occurrence of the top event. Additionally, the FTA can provide the relative importance of the basic events, and the sensitivity of the top event to the different basic events. With this knowledge, actions and resources can be prioritized to achieve the biggest reduction in the likelihood of occurrence of the top event.

One of the strengths of the FTA is that unlike the Failure Modes and Effects Analysis (FMEA), for which architectural hierarchy is material and relevant to the analysis, in the FTA hierarchy doesn't matter. The FTA simply looks for the contributors to an event and crosses the hierarchical boundaries. Another strength of the FTA is that it is easy to learn, and easy to understand.

At the service of risk management, the FTA addresses only the pathways to Hazardous Situations. There may be many more top events that are of interest to, e.g., reliability, or others. Also, by its nature, the FTA only detects the events that are found in the path to the Hazardous Situation. Moreover, the modeled faults are not exhaustive—only the faults that are considered relevant and credible are assessed.

In the FTA the words fault and failure are meaningful depending on the context in which they are considered. Failures can be the result of faults. But when a fault is viewed in consideration of *its* underlying contributors, then it can be seen as a failure. The word "event" is a more general term that can be applied to both faults and failures.

12.1.2 Theory

An FT is a graphical representation of parallel and sequential events that are interconnected by logic gates, leading up to a top event. The top event usually represents an undesired outcome, such as a Hazardous Situation, and the lower events include faults, user errors, and normal conditions. An example of an FT can be seen in Fig. 12.1. The logic gates show the required relationships among the lower level events that are needed to cause the output of the gate in question. The event at the top of a gate is called the "higher" event and is the output of the gate. The events below a gate are called the "lower" events and are the inputs to the gate.

Due to its logical construct, an FT can always be translated into a set of Boolean equations. As such, rules of Boolean algebra can be applied to FTs to simplify and reduce them. This simplification is beneficial both in understanding of the System under analysis, and also for the derivation of minimal cut sets of the tree.

Analysis of FTs can provide us with:

- minimal cut sets of the tree
- qualitative component importance
- knowledge of cut sets that are susceptible to CCFs

If probabilities of basic events are known, the quantitative analysis can provide probability of occurrence of the top event; quantitative importance of the minimal cut

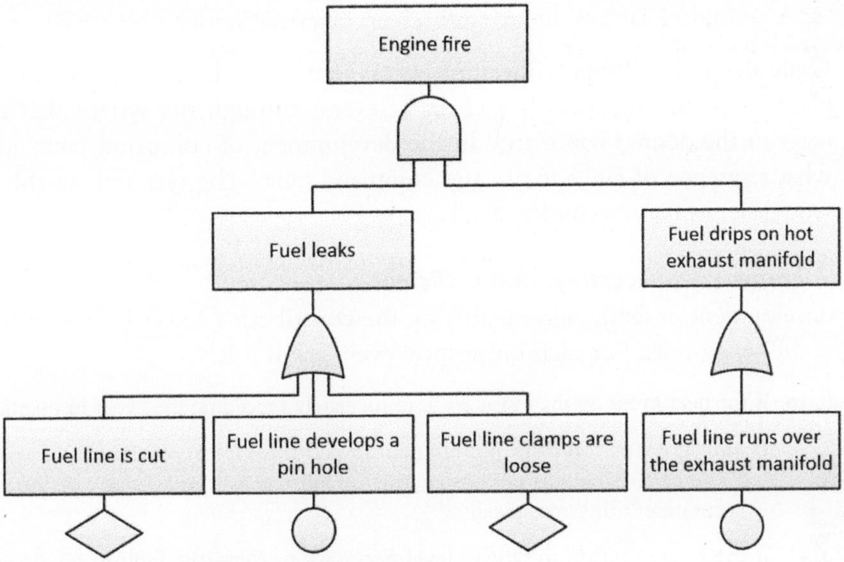

Figure 12.1 Example of Fault Tree Analysis diagram.

sets and components; and sensitivity evaluations. Sensitivity evaluation shows the sensitivity of the top event to the variations in probabilities of occurrence of basic events that lead up to the top event.

12.1.2.1 Primary, secondary, and command Faults

Ref. [24] identifies three categories of faults: primary, secondary, and command, and describes them as follows:

- *Primary fault*: "Any fault of a component that occurs in an environment for which the component is qualified"; e.g., "a pressure tank, designed to withstand pressures up to and including a pressure P_0, ruptures at some pressure $p \leq P_0$ because of a defective weld."

  A primary fault is inherent to the component and occurs under design conditions. A primary fault in one component is presumed to be independent of a primary fault in another component, i.e., a primary fault in component X would not cause a primary fault in component Y.
- *Secondary fault*: "Any fault of a component that occurs in an environment for which it has not been qualified. In other words, the component fails in a situation which exceeds the conditions for which it was designed; e.g., a pressure tank, designed to withstand pressure up to and including a pressure P_0, ruptures under a pressure $p > P_0$."

  A secondary fault is due to unforeseen external factors.
- *Command fault*: "Proper operation of a command, but at the wrong time/place." A command path is a chain of events through the system that culminates in the desired command. In the development of command-faults identify what sequence of faults led to the command fault. The terminus of this chain would be primary/secondary faults.

12.1.2.2 Immediate, necessary, and sufficient

In the development of faults, and identifying the contributing lower level events, consider the following tests. For each lower level event ask if it is:

Immediate Is the next event on the lower level, immediately preceding the event in question?

Benefit: It keeps you from jumping ahead and missing causes. Ref. [24] says "The basic paradigm in constructing a fault tree is 'think small', or more accurately 'think myopically.'"

Necessary Is the next event on the lower level necessary to cause the fault in question?

Benefit: It helps maintain the cause and effect relationship, and avoids extraneous entries.

Sufficient Do you have *all* the necessary events to cause the fault in question?

Benefit: Ensures the higher event can actually happen, given the cited lower-level events.

For the purposes of risk management, we can presume that the design has already been peer reviewed, analyzed, modeled, and tested. Therefore faults due to designer errors can be excluded.

12.1.2.3 State of Component–State of System

A fault can be a State of Component, or State of System. Ref. [24] defines "A 'state of component' fault is a fault is associated uniquely with one component. A 'state of system' fault is a fault not uniquely associated with one component. The immediate cause of a 'state of system' fault involves more than one component."

12.1.2.4 Common Cause Failures

CCFs describe situations where two or more component failures occur due to a common cause. In the paradigm of immediate-necessary-sufficient logic, common causes of the failures are not explicitly modeled in the FT. Instead, common causes describe an implicit dependency where multiple faults are triggered when the common cause occurs. Once the CCFs are identified, they should be included in the FT to raise awareness to their contribution.

Examples of CCF are as follows:

- Environmental factors, e.g., temperature, humidity, pressure, and vibration
- Faulty calibration
- Error in manufacturing, causing all copies of a product to be faulty
- Erroneous work instruction causing all users to operate the device incorrectly

CCFs are particularly important in redundant systems, where safety is based on the presumption of unlikelihood of simultaneous failure of the redundant systems. In such cases, a single cause would simultaneously defeat all of the redundant systems.

12.1.3 Symbols

FTs are constructed from logical connection of nodes. The nodes represent various events such as a basic event, an undeveloped event, etc. The nodes are connected to each other using logic gates. Special symbols are used to represent different types of nodes and gates. Some of the most common symbols are presented in Fig. 12.2.

For convenience in labeling the events and gates, sometimes the style in Fig. 12.3 is used, where the descriptive text is entered in the rectangles above the FT symbol.

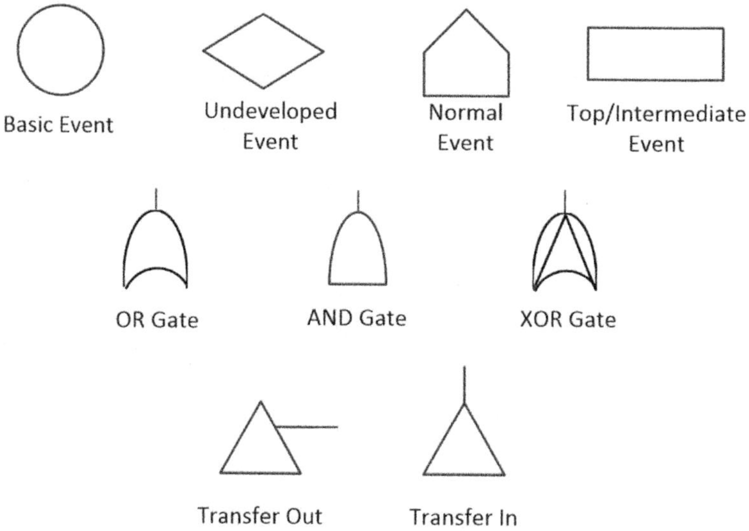

Figure 12.2 Fault Tree Analysis symbols.

Definitions of FTA symbols

Basic Event	A basic initiating event, e.g., component faults, requiring no further development
Undeveloped Event	An event that is not further developed due to lack of information, or when the consequences are not important
Normal Event	An event that is normally expected to occur, e.g., the device gets used
Top/Intermediate Event	An event that is further analyzed
OR Gate	Output occurs when one or more of the inputs occur
AND Gate	Output occurs when all of the inputs occur
XOR Gate	Output occurs when only one of the inputs occurs
Transfer	A symbol that shows part of the tree is transferred to another location. Used to manage the size of the tree on a page, and to avoid duplications

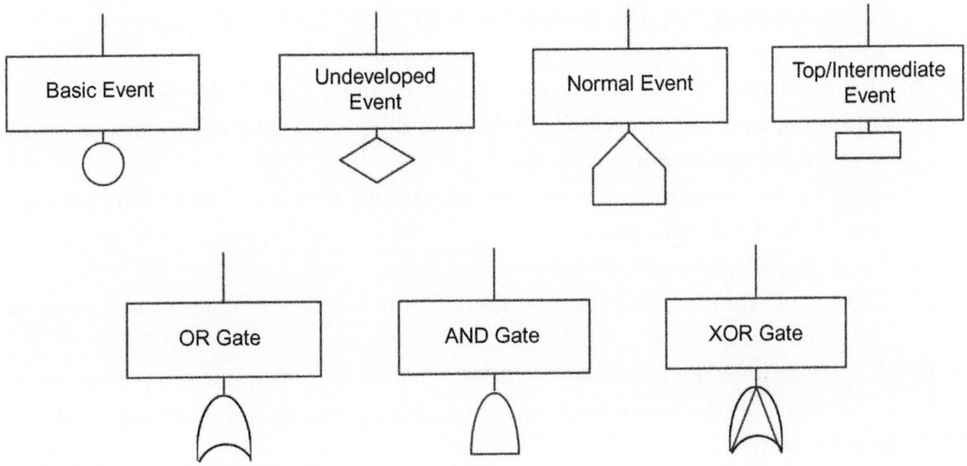

Figure 12.3 Alternate Fault Tree Analysis symbols.

12.1.4 Methodology

1. Obtain the System requirements and architectural design. Understand the theory of System operation.
2. Define the boundary of analysis. Know what is in, and what is out. "What is in the analysis will be those contributors and events whose relationship to the top undesired event will be analyzed. What is out of the analysis will be those contributors that are not analyzed" [24]. External interfaces to the boundary of analysis should be included in the analysis as influencers. However, we don't analyze the causes of the behavior of the external influencers. It is also possible to make assumptions about the external influencers. For example, you may assume for a mains-powered device, that mains power frequency will be 60 Hz.
3. Define the top Events. In risk management these would be Hazardous Situations. You could also use Harms as the top events. But in the construct of the BXM method, Harm is presumed in the face of a Hazardous Situation (see Section 3.2).

   The System context, or initial state, may need to be specified for the top events.
4. For each fault, check to see if it is a State-of-Component (SC) fault. If not, it is a State-of-System (SS) fault (see Section 12.1.2.3 for more details). Tag the fault with SC/SS. If it is a SS fault, develop it further. If it is a SC fault, it is caused by primary faults, secondary faults, or command faults (see Section 12.1.2.1 for more details). If two or more lower level faults contribute to the component fault, use an OR gate to flow the lower level faults to the component fault.
5. Develop each fault—ask what are the immediate, necessary, and sufficient lower level events to cause the higher level fault (see Section 12.1.2.2 for

more details). Using the appropriate logic gates connect the higher level fault to the lower level events.

6. Repeat Steps (4) and (5) for every fault in the FT until the terminus of the tree is reached. The terminus is where all the events are basic events, or undeveloped events, or normal events.
7. Identify the components that are susceptible to CCFs, and then properly model the CCF contribution.
8. Look for dependencies of faults.
9. Determine the minimal cut sets. Identify any minimal cut sets that depend on a singular basic event.
10. If quantitative data is available for the basic events, compute the probability of occurrence of the Hazardous Situations. Also perform an importance and sensitivity analysis.

Information from the FT can be used in the Preliminary Hazard Analysis (PHA; Section 12.3). For example, probability of Hazardous Situations would inform the P_1 values; and sequence of events can be created from the causal chains in the FTs.

Passive versus Active components. Passive components' contribution to the system is more or less static. Examples are: wires, tubes, and welds. Active components provide a dynamic contribution to the system. Examples of active components are valves and switches. Historically, from a reliability perspective, passive components are two to three orders of magnitude more reliable than active components.

Tip – Because passive components are far less likely to fail than active components, you may want to exclude passive components from the fault tree analysis, as their contribution to System risk will be small.

It is important to develop a FT to a sufficient depth to gain meaningful knowledge of failure mechanisms and functional/failure dependencies. Developing FTs beyond that is a waste of effort, and potentially distracting. Another drawback to overly deep FTs is that generally quantitative failure rate data doesn't exist for low-level components.

A common heuristic is to model the system to the depth necessary to identify functional dependencies, and to a level for which failure rate data exists for the components.

12.1.5 Ground rules

The ground rules listed in this section are intended to facilitate the creation of FTs while minimizing confusion and wasted effort.

12.1.5.1 Write faults as faults

Choose the appropriate syntax. State what the fault is, and if conditions of the fault are material, state under what conditions. You may need to be verbose. Write it in a way that in the future, another person or even yourself can make sense of the fault description. Example: catheter balloon bursts when inflated by the surgeon.

12.1.5.2 No gate-to-gate connections

Gate inputs should be properly defined. A direct gate-to-gate connection is a shortcut which bypasses defining the lower level gate's output. While it may be tempting to take such shortcuts, it makes the FT more difficult to read and understand by others.

12.1.5.3 Mark low-likelihood faults as Basic Events

If it is clear that a fault is of very low likelihood, do not further develop it to lower levels. Mark it as a Basic Event.

12.1.5.4 Don't model passive components

Components are either passive, e.g., a wire, or a pipe. Or, active, e.g., a switch or a valve. Historically, it's known that passive components failure rate is 2–3 orders of magnitude smaller than active components. While it is technically possible to model passive component faults, it does not add much value to the risk analysis of a product.

12.1.5.5 Be judicious in modeling secondary faults

Since the purpose of the FTA is identification of credible Hazardous Situations, be cautious about modeling secondary faults. Secondary faults occur under out-of-design conditions. Out-of-design conditions are usually unlikely. Consider the added value of modeling such faults.

Tip – Use a skilled facilitator to guide the FTA work sessions. Good facilitation guides the brainstorming, and proper fault tree construction, and also prevents confusion of participants on concepts such as Hazard, Causes, and Basic Events.

12.2 MIND MAP ANALYSIS

12.2.1 Introduction

The use of diagrams to graphically map information is a very old technique. The term "Mind Map" was first popularized by the British pop psychology author Tony Buzan. A mind map is basically a tool for graphically organizing thoughts and ideas. It is a very useful tool in brain storming, and is an accessible alternative to FTA (see Fig. 12.4 for a simple example).

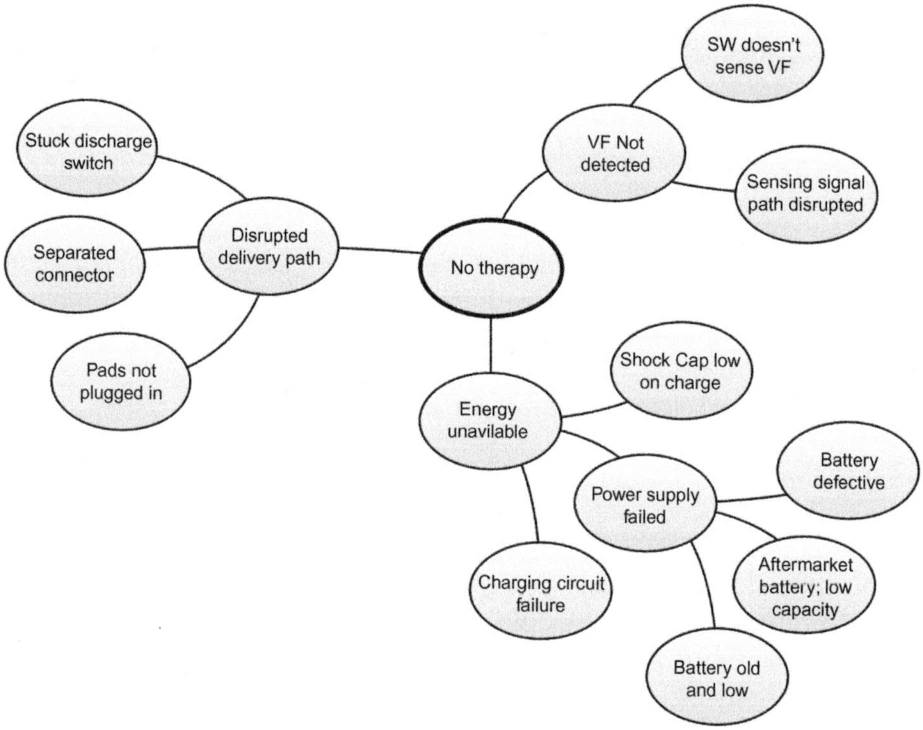

Figure 12.4 Example of Mind Map.

An advantage of Mind Map over FTA is that it is simpler to learn than FTA and software for Mind Mapping is either free, or very low cost.

12.2.2 Theory

Starting from a central theme, which is marked as the central node, second layer nodes are connected to the central node. The second layer nodes comprise all the immediate, necessary, and sufficient pathways to the central node. This pattern repeats for each second layer node, and further nodes successively.

In the absence of FTA's specialized nodes such as Normal Event, and Undeveloped Event, develop each branch to the degree that it makes sense. If an event is undeveloped due to the current lack of information, you can put a TBD in there as a reminder to come back to it.

The main purpose for constructing a Mind Map is to graphically tell the story of how an undesired event can happen. This information will later be captured in the Sequence of Events in the PHA.

Unlike the FTA, Mind Maps do not use logic gates. Implicitly all connections are OR gates. For example, in Figure 12.4 under "VF not detected," it can happen if either "SW doesn't detect VF" OR "Sensing signal path is disrupted."

12.2.3 Methodology

Step 1. Obtain the System requirements and architectural design. Understand the theory of System operation.

Step 2. List the top undesired events for the System. These could be the Harms, or the Hazardous Situations.

Step 3. Define the scope of analysis. What's in; what's out.

Step 4. For each top undesired event create a Mind Map. Brainstorm and identify pathways to the top event.

Tip – Although a Mind Map doesn't model logic gates and the connections are interpreted as OR gates, a work-around would be to create a node called "AND" where multiple branches feed into it. The output of the AND node would occur if all the incoming branches occur.

12.3 PRELIMINARY HAZARD ANALYSIS

12.3.1 Introduction

PHA is a technique that can be used early in the development process to identify the Hazards, Hazardous Situations, and events that can cause Harm when few of the details of the medical device design are known. The PHA can often be a precursor to further studies.

With the advance knowledge that is generated by the PHA, it becomes possible to identify the safety-critical parts of the System concept, estimate the potential risks associated with the System, and thus guide the design team to prioritize and focus resources on the highest risk parts of the System.

Performance of a PHA is far more effective when people from various functions are engaged to participate. This provides for a multiperspective analysis leveraging insights from many points of view. Here, risk management can serve as a tool for stimulation of communication among the team members who would ordinarily not have reason to communicate.

R&D engineering benefits from the advance knowledge generated by the PHA to anticipate Risk Controls and design them into the System early, instead of late in the product development process and thus reduce product development costs.

Another significant benefit of the PHA is that it can provide advisory to management *not* to proceed with the development of a product, in case it is anticipated that the risks of the yet undeveloped device will outweigh its benefits. It is far less expensive to cancel a project early, than to embark on the project and have to cancel it late in the design and development phase.

Basically, the PHA is an early version of the Hazard Analysis Report (HAR). In fact, it uses the same Risk Assessment and Control Table (RACT) template. The main difference is that the PHA has access to little actual design information and uses a lot of estimations. Also, the PHA is not a living document. It is only used as guidance to start the project, and also as a reference for the HAR. But it is the HAR that lives on for the life of the product.

It is highly recommended that a PHA be performed especially for new and novel product development.

12.3.2 Methodology

Inputs to the PHA are:

- System requirements
- Concept architecture
- Intended use, intended user, and intended use environment
- Risk acceptability criteria (from the Risk Management Plan (RMP))
- Clinical Hazards List (CHL)
- Harms Assessment List (HAL)

If the System is a new version or an iteration of a product that has been previously analyzed for risk, then it is advisable that the previous risk analysis be used as an input to the new analysis.

In the following sections the PHA workflow is described.

12.3.2.1 Safety characteristics

Given the intended use, intended user, and intended use environment for the System, and given the concept architecture and System requirements, identify those qualitative and quantitative characteristics that could affect the safety of the medical device. Consider the technologies used in the device, and how they can contribute to Hazards. Where appropriate identify the operational specification limits within which the device can be operated safely.

Include user interface (UI) characteristics that could be related to safety.

Annex C of ISO 14971 [3,7] provides questions that can be useful aids in this effort.

12.3.2.2 Identify System Hazards

Using the knowledge gained in Section 12.3.2.1 examine the CHL. Identify the Hazards that are relevant to the System. Exclude the remaining Hazards in the CHL from the analysis and provide rationales for why any Hazard is excluded.

Also, consider potential Hazards that could be encountered under reasonably foreseeable misuse conditions. See Section 3.1.1.1 for a definition of "reasonably foreseeable misuse."

Interfaces are a common source of failures. Pay particular attention to interfaces to parts of the Systems that are designed by an external entity (supplier). Remember to consider potential Hazards arising from interfaces between use-conditions. For example, the maintenance function is not commonly in the forefront of the mind of medical device designers. Could a maintenance person leave the device in a potentially unsafe state for the clinical user?

Postulate the potential Hazardous Situations for the System, and perform an FTA to determine pathways that could lead to the identified Hazardous Situations (see Section 12.1 for instructions on how to perform an FTA).

Populate a RACT template with information from the FTA. All applicable Hazards from the CHL should appear in the RACT. Each Hazard will have one or more Hazardous Situations associated with it. Document the potential pathways to each Hazardous Situation. These pathways are easily derived from the FTA. Fill in a P_1 value for each pathway. P_1 is the probability of occurrence of the Hazardous Situation. If field data is available for P_1, use that. Otherwise, using the collective team judgment estimate a value for P_1. When estimating P_1, use a basis that makes sense for your product, and facilitates your Risk Control decision making. For a long-term implantable product, a unit such as *patient-year* is suitable. But for a device that is used repeatedly, *per-use* makes more sense. For example, P_1 for the Hazardous Situation of "Over-infusion of insulin" due to an erroneous blood glucose reading could be stated as "10^{-3} per use."

Tip – Estimate P_1 for one device, not the entire fleet of devices. The reason is that making more devices doesn't make any single device less safe. If P_1 is estimated over the whole fleet of devices, the risk for a given Hazard would appear to continuously grow as more copies of the device are sold.

The next step is to evaluate the Hazardous Situations and identify the potential Harms that could ensue from each of them. In the BXM method, for every Harm there are five P_2 numbers, which are the probabilities of sustaining Harm in different severity classes (see Section 11.6 for more information on the HAL). Look up the P_2 probabilities in the HAL and populate the RACT table with their respective values. If your method uses a single P_2 number, then just use that.

Compute the risks for each Hazardous Situation by multiplying the P_1 for each Hazardous Situation by the five P_2 numbers from the Harms. This will result in five risk-numbers—one for each severity class.

Compute the residual risks for each Hazardous Situation, and also for the overall System (see Section 17.3 for details on how to do the computation).

At this point the PHA is ready to serve its purpose and answer the following questions:

1. Can the System be built such that its risks are acceptable?

   This is to advise the management on whether they should commit resources to design and development of the medical device. Using the risk acceptability criteria in the RMP evaluate the overall residual risks of the System. If the risks in all severity classes are acceptable, the answer to the question will be affirmative. But if the risks in one or more of the severity classes come out to be unacceptable, then investigate with the RMT, and R&D to determine whether the team believes that with additional future Risk Controls, they can feasibly bring the System risk down low enough so that in the end, the overall risks of the medical device become acceptable. If they can, then again the answer to the question is affirmative. Else, the answer to the question is negative and the project should not be started.

   Some factors may give you early warning of potential future problems. For example, if the concept for an implantable device requires the use of a toxic metal such as nickel or mercury, you could anticipate the possibility that in the end the residual risk of the product may be unacceptable.

2. What are the most safety-critical aspects of the System?

   The answer to this question helps to focus resources on the most important safety-critical aspects of the System. Look for any Hazardous Situation which has a risk in the unacceptable zone. They should be the highest priority areas for the Design and Development tram. If all the risks of all the Hazardous Situations are acceptable, then make a subjective judgment on how to prioritize them. For example, you could see how close any Hazardous Situation is to the unacceptability boundary, and prioritize by the distance to unacceptability.

12.4 FAILURE MODES AND EFFECTS ANALYSIS

FMEA is a systematic method of exploring the ways in which an item or process might potentially fail to achieve its objective, and the effects of such failures on: the performance of the system, or process, or the environment and personnel. FMEA is a forward reasoning process, also referred to as bottom-up, or inductive analysis. The FMEA technique was originally developed by the US military in 1949 as a reliability analysis tool. Later, it was used by NASA in many space programs. Today, many industries, in particular the automotive industry, use this analytical tool to improve the quality of their products.

There are different types of FMEA processes serving different purposes. The BXM method adapts the FMEA for the benefit of medical device risk management and uses four types of FMEA: Design Failure Modes and Effects Analysis (DFMEA), Software Failure Modes and Effects Analysis (SFMEA), Process Failure Modes and Effects Analysis (PFMEA) and Use-Misuse Failure Modes and Effects Analysis (UMFMEA). At the service risk management, FMEAs are to identify Hazards, and estimate the likelihood of their occurrence.

It is important to distinguish two terms: Fault and Failure. A fault is an anomalous condition for a part. A failure is the inability of an entity to achieve its purpose.

- A fault *could* result in a failure, but not necessarily
- A failure may occur with no faults

With respect to risk management, FMEAs are used to identify Failure Modes which can result in Hazards or Hazardous Situations. It is important to realize that occurrence of faults and failures can result in Hazards, but not necessarily. And Hazards or Hazardous Situations can occur in the absence of any fault/failure. To elucidate—a medical device that is designed for adults, if used on children may create a Hazardous Situation, even though the device is working perfectly according to its design. Or, a medical device may have a fault that doesn't create a Hazard.

In the FMEA, the subject of analysis is decomposed into elements. The granularity of this decomposition is subjective and is called the level of indenture. During the course of the analysis, the Failure Modes of each element and consequences on the subject of analysis is considered. In general, the identification of Failure Modes and the resulting effects is based on experience with similar products and processes, or on knowledge of the applicable science.

The subject of the FMEA analysis may be the entire medical device, a subsystem, a component, a process, or anything that the analyst chooses.

Tip – When choosing the granularity of decomposition of the subject of FMEA (level of indenture) save time and resources by not going deep into the parts of the System that are understood and have a well-known history.

It is important that the scope of analysis be clearly defined and understood. That is, the boundary of analysis, and what is included in the analysis should be clearly defined. Interfaces to the subject of the analysis should also be clearly identified.

Context of operation is important. The same Failure Mode could have starkly different severities, depending on the context of operation. For example, the failure of a jet engine on an aircraft has very different consequences whether the aircraft is in flight, or on the ground.

12.4.1 Facilitation of FMEAs

Successful execution of FMEAs requires skillful facilitation of the FMEA sessions. The facilitator takes responsibility to:

- Convene the FMEA work sessions
- Ensure persons with appropriate competences are present
- Ensure proper resources, materials, and support elements are present at the working sessions
- Ensure the purpose of the FMEA, the ground rules, and context of operation are understood at the start of each working session
- Assist in the discernment of Causes, Failure Modes, effects, and mitigations
- Limit lengthy discussions and guide the team to useful conclusions
- Help the team maintain focus, and remain within the scope of analysis
- Manage the working sessions and terminate when productivity of the team declines due to fatigue, or other factors.

12.4.2 Hierarchical multilevel FMEA

Hierarchical multilevel FMEAs is a strategy that you can adopt to enable efficiencies of parallel processing, and modular reusable analysis. This technique requires a system decomposition based on the system architecture as explained in Section 10.1.

The System, referred to as Level 1, or L1 in Fig. 12.5, is decomposed into two Level 2 (L2) components. Each L2 component is further decomposed into multiple L3 components and so on. You can perform DFMEAs and PFMEAs on the lower level components. And just as the system itself is progressively integrated from lower levels up to the top level, so too can the lower level FMEAs be progressively integrated until the top-level System DFMEA is generated.

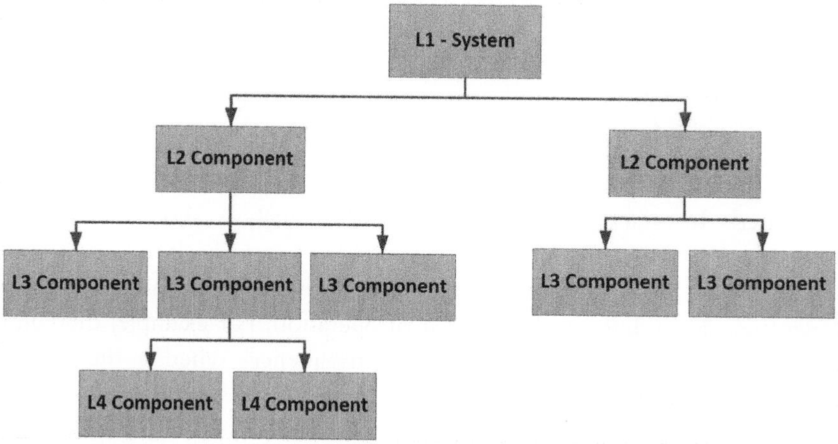

Figure 12.5 Multilevel hierarchy.

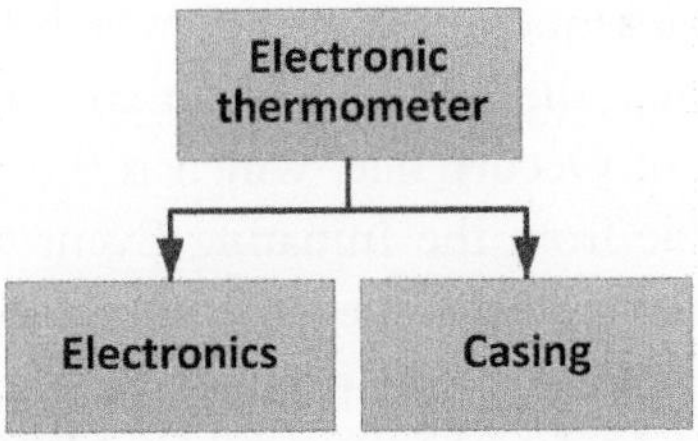

Figure 12.6 Electronic thermometer.

The benefits of this technique are multifold:

1. Parallel work—Let's consider the example in Fig. 12.6, which describes an electronic thermometer decomposed into the electronics and the mechanical casing. It would be possible for the electronic team to be working on the FMEA of the electronics, while the mechanical team works on the casing FMEA. Later, the results of their work are integrated in the DFMEA of the Electronic Thermometer System.
2. Modularity and reuse—If in the example of Fig. 12.6 the manufacturer decides to update the thermometer casing for a fresh look, but keeps the electronics the same, the hierarchical multilevel FMEA technique allows the reuse of the electronic DFMEA, while refreshing only the Casing DFMEA/PFMEA.

Specifically, the *DFMEAs* lend themselves to modular reusability. PFMEAs can be reused in the ordinary sense of basing a new analysis on old work and making revisions thereto. UMFMEAs are unique to each System and any reuse depends on the degree of similarity between the current System and the previous System.

For DFMEAs to be modularly reusable, strict adherence to the scope, and boundary of analysis is required. In the DFMEAs, for Failure Modes that have a safety impact, Severity and Detectability ratings depend on the context of use. Therefore during reuse of DFMEAs, for Failure Modes with safety impact, Severity and Detectability ratings need to be reexamined within the context of use.

For more details on the integration of DFMEAs see Section 16.1.

12.4.3 Failure theory

In the context of FMEA the End Effect is the outcome of a chain of events. Fig. 12.7 shows a model of this concept.

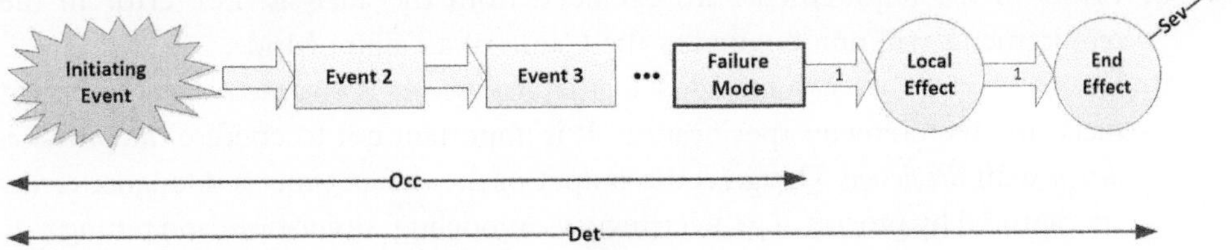

Figure 12.7 Failure theory.

An initiating event starts a sequence of events, which leads into the Failure Mode. Once the Failure Mode occurs, the Local Effect (if any) and the End Effect will happen. Therefore the concept of Occurrence, which is the probability of occurrence of the Failure Mode is applicable from the Initiating Event to the Failure Mode. Occ is the probability of occurrence of *all* the events in the causal chain. Occurrence rating is inclusive of the implementation of all pertinent mitigations. The probabilities of occurrence of the Failure Mode, Local Effect and End Effect are the same.

Severity is the property of the End Effect. Severity is the significance of the worst reasonable consequence of the End Effect at the boundary of analysis.

Detectability is applicable to the entire chain of events, from the Initiating Event to the End Effect. Detection may happen when the initiating event happens; or somewhere along the chain of events; or even after the End Effect has been manifested. In detection, there is an implicit assumption that countermeasures are feasible to reduce the Occ or Sev ratings.

12.4.4 Ground rules

Ground rules are a set of understandings and agreements that the FMEA team uses to ensure smooth and productive work sessions. Ground rules can be expanded, refined, or clarified as the process continues.

Below are a set of suggested ground rules. You may adapt and adopt them as you see fit for your purposes.

1. Only one failure is considered at a time.
2. The function of each item under analysis must be clearly known and stated. An ambiguous statement of function makes it difficult to tell whether the item has failed.
3. Context of operation shall be stipulated.
4. Failure shall be defined. In some cases, it may not be clear how much degradation in the performance of an item would constitute a failure.
5. Only reasonable Causes and Failure Modes are considered.
6. If a failure results in multiple End Effects, each End Effect is listed in a separate row.
7. If a Failure Mode can be caused by different causal chains, each causal chain is listed in a separate row.
8. Errors in the requirements are excluded from the analysis, i.e., error in the requirements will not be cited as the Cause of a Failure Mode.
9. Designer errors are not included in the analysis. It is assumed that the design meets the requirements specification. It is important not to confuse the *process of design* with *the design*. Design is the output of the design process. Designer errors are captured by process, e.g., peer reviews, modeling, simulation, and testing.

10. If a mitigation eliminates a Failure Mode, or makes its likelihood indistinguishable from zero, you may delete that row, or keep that row in the FMEA as historical information for the benefit of future readers/users of the FMEA. If it is decided to keep the row, clearly mark it as not credible, and for informational purposes only.
11. In order to maintain focus of FMEAs, DFMEA will assume that manufacturing is correct; PFMEA will assume that design is correct.
12. From the risk management perspective, it may be tolerable to have a high-criticality Failure Mode remain in the FMEA, if the Hazard from the End Effect of that Failure Mode is mitigated elsewhere in the System, such that the patient is kept safe from that Failure Mode.
13. The FMEAs in the hierarchical multilevel structure will use the same methodology and scales for rankings. This is to enable and facilitate integration of the FMEAs.

As stated above, these ground rules are intended to make the FMEA sessions flowing and productive. If, e.g., analysis of a Failure Mode reveals a missing requirement, or a design error, it doesn't mean you have to ignore it. To the contrary! You should communicate that to the product development team. This is how FMEAs add value to product development process.

12.4.5 On merits of RPN for criticality ranking

The RPN method is a common and historical practice which uses the product of Severity, Occurrence, and Detectability, $S \times O \times D$ as a means to prioritize the Failure Modes by criticality. Higher RPN indicates higher criticality. This is an easy to understand and implement technique. But there are many drawbacks with the RPN method. Namely,

- *RPN is not continuous.* In a scale where *S*, *O*, and *D* are ranked in 5 ordinal grades, the RPN range is 1–125. But many of the numbers in this range never manifest. For example: 28, 31, 49, etc.
- *RPN sensitivity to other factors.* Consider a Failure Mode whose severity is 5 with one whose severity is 4. If $O = 2$, and $D = 1$, the RPNs will be 10 versus 8—a difference of 2. But if $O = 4$, and $D = 3$, the RPN would be 60 versus 48—a difference of 12.
- *Consecutive ordinal numbers are not linearly spaced.* For example, in Occurrence ratings, the difference between 5 and 4 is usually not the same as the difference between 4 and 3. It may be that a logarithmic scale is used for Occurrence ratings, e.g., 10^{-3}, 10^{-4}, 10^{-5} in which case the difference between adjacent ranks is a factor of 10.

Table 12.1 Example criticality matrix

Criticality		Severity				
		1	2	3	4	5
Occurrence	1	2	2	3	3	3
	2	1	2	2	3	3
	3	1	1	2	2	3
	4	1	1	1	2	3
	5	1	1	1	2	2

There have been attempts to improve the RPN method. One such method is called ARPN, where the Severity, Occurrence, and Detectability are placed on a logarithmic scale, and thus instead of multiplying the *S*, *O*, and *D*, they are added together.

The BXM method suggests a modified Pareto principle in the use of RPNs. Sort the Failure Modes by descending RPN values. Take the top 20 ± 10% of the Failure Modes. The reason for ± 10% is that usually there is a natural break around 20%. The ± 10% gives the flexibility to *find* and use that natural break-point. In addition to these top ranked Failure Modes, include any Failure Mode with a Severity ranking of ≥ 4 in the high-priority group for mitigations.

As stated earlier, RPN is a simple, but coarse method for prioritization of criticalities. A finer method is using a criticality matrix such as Table 12.1 to customize the criticality rankings to suit your organization. There is no reason why the matrix should be two dimensional. Addition of a third factor, e.g., Detectability, would make the matrix three dimensional. While graphical display of a 3-D matrix on a page may be less convenient, computers have no problem resolving and ranking criticalities based on your design of the criticality matrix.

12.4.6 Benefits of FMEA

Although in this book FMEAs are used for the benefit of safety risk management, there are many other advantages and benefit to FMEAs. By its nature, FMEA is systematic and exhaustive. It examines *every* element in the scope of analysis for their Failure Modes, and effects. This helps with the detection and elimination of product Failure Modes, thus improve product reliability and quality, which in turn should improve customer satisfaction.

FMEA is a predictive analytical tool. It enables early identification and handling of Failure Modes, thus reducing product development costs, by avoiding late-stage changes and corrections.

Another benefit of FMEAs is the discovery of missing or wrong requirements. Designers design devices per the System requirements. Performing an FMEA on the design could reveal that although the design meets the requirements, certain Failure Modes' criticality ratings are too high. Thereby feedback is given to design engineering to revise the requirements, or even add requirements.

In safety risk management, FMEA is used as a tool for identification of Hazards and the sequences of events that could lead into those Hazards. Also, Occurrence ratings of the Failure Modes are used in System risk estimation. FMEAs detect many Failure Modes, some of which have no impact on safety. For example, failure of an electronic thermometer to power up is a reliability issue that could be annoying to a nurse, but is not safety critical. Such findings have impact on business, customer satisfaction, etc. but do not create Hazards. It is beneficial to the business to know *all* Failure Modes of the System—safety-related or not. While safety-related Failure Modes should generally be mitigated as far as possible, the decision on whether, and how much to mitigate nonsafety-related Failure Modes is entirely a business decision.

To get the most out of FMEAs, start FMEAs early and iterate. You can start as soon as a concept and block diagrams are available and do high level functional analyses. As more details of design becomes available, update the FMEAs and continue to iterate until the end of the design process. This diligence pays off by not only helping the designers with discovery of weaknesses in their design, and rectifying them before the product gets into the field, but also after the product is released, any proposed new changes can be easily evaluated for impact to safety.

12.4.7 FMEA weaknesses

Despite its power and utility, the FMEA technique has some weaknesses. Some of these weaknesses are as follows:

- The FMEA is unable to detect End Effects that require multiple Failure Modes.
- Because the FMEA fundamentally treats each Failure Mode individually, as an independent failure, CCF analysis is not well suited to the FMEA model.
- The FMEA doesn't catch hazardous End Effects that are not due to failures, e.g., Hazards due to timing issues, physiological variability, etc.
- Performance of FMEAs is time consuming.
- FMEA is difficult to master.

With the knowledge of the strengths and weaknesses of the FMEA as a tool, you can properly benefit from the value that it offers without being blindsided by its shortcomings.

12.4.8 Ownership of FMEA

It is recommended that design engineering own the DFMEA, SFMEA, and UMFMEA, and manufacturing engineering own the PFMEA. Ownership means taking responsibility for the creation and maintenance of the work products. Some of the reasons for this preference are:

1. Mitigations are the domain of design and manufacturing engineering. They can best determine the possibility, practicality, and impact of mitigations on design and manufacturing.
2. Risk management is only one of the beneficiaries of the FMEAs and uses the FMEAs as a tool for detection of the System Hazards. Since risk management is focused on safety, if risk management owns the FMEAs, the attention would be primarily on the safety-related Failure Modes and some of the nonsafety-related Failure Modes may not get the attention that they deserve. Therefore the knowledge that could be gained and the value that could be delivered to product development may not become realized.

Tip — Involve the FMEA reviewers and stakeholders in the production of the FMEAs. Not only the collective participation enriches the analysis, but also the familiarity which is gained as a result of the participation, will make the review of the FMEAs easier.

12.4.9 Making your way through the FMEA

Performing FMEAs on any product of moderate to high complexity takes a large amount of time and resources. Often participants get tired and the quality of their input declines. You could even witness lengthy arguments that don't come to any conclusions. This is one of the reasons people tend to shy away from doing FMEAs. Or, if they do it, they try to get through it as quickly as they can, and check the box as "done."

Here are some of the causes for unsuccessful FMEA sessions:

- The team is sequestered for long sessions for several days.
- The team loses focus on what is the subject of the analysis, or what is the context of operation for the subject of the analysis.
- Only a small part of the FMEA spreadsheet is projected on a screen; people can't see all the columns, or column headings; they get lost.
- One person dominates the conversation; others quiet down and just nod in agreement.
- Participants check their emails, smart phones, or do other work and lose focus.
- People get confused and have trouble distinguishing Causes from Failure Modes, and effects.
- The team is scoring severity and occurrences as they go; this promotes inconsistencies in ratings as people's frames of minds drift over time.

- People get stuck in long discussions on some items, which causes some participants' attention to drift off.
- People get tired, creativity ceases, and generic vague answers are put into the analysis.
- The product design process is already finished. There is no opportunity to make any changes. Findings of high significance will cause great pains and costs for the business. Therefore there is reluctance to do any deep analysis in the fears of finding something significant.
- Granularity in rating scales is too high. When estimating a rating, a lower granularity scale is better than a higher granularity scale. A high granularity scale, e.g., a 10-point scale, could lead into unnecessary long debates in choosing between, e.g., a 6 versus a 7 rating because the differentiation between adjacent ratings may not be so clear.

Some tips to help smooth the FMEA process and ensure success are as follows:

- First and foremost, use a skilled facilitator for the FMEA. Section 12.4.1 describes some of the responsibilities of an FMEA facilitator. Most FMEA participants participate in FMEA work only occasionally. Therefore they become rusty on the mechanics of the analysis. With coaching and guidance from the facilitator, most participants climb the learning/remembering curve quickly.
- Where possible, try to reuse existing FMEAs to accelerate the work.
- Keep the duration of the sessions to less than 3 hours. Long sessions lead to fatigue and reduced quality of work.
- Refresh participants on the ground rules, definitions of Failure Mode, Local and End Effects, and be vigilant to ensure the entries in the FMEA are properly worded.
- If necessary, give a quick overview of failure theory, and definitions of Severity, Occurrence, and Detectability.
- Make the definitions of the rankings for Severity, Occurrence, and Detectability easily accessible, e.g., by printing them on posters and posting them on walls.
- Make sure the agenda and objective of the meeting is clear, state it at the beginning of the meeting and post it on a wall.
- Prework—participant should become familiar with the design under analysis, before coming to the meeting. The FMEA session time is precious and should not be used for explaining the basic understanding of the subject of analysis.
- Have physical samples, models, drawings in the room that participants touch and use as discussion tools. Simply touching a physical sample is a great thought stimulator. Also, it is much easier to convey thoughts and ideas about Failure Modes using models/drawings.

- To improve consistency of ratings throughout the FMEA finish the brainstorming then identify mitigations, and then begin the ratings.
- Control the conversation and maintain focus; avoid sidebar conversations, lengthy war stories, and egotistical debates.
- During the brainstorming part, try to maintain momentum by just capturing the ideas and tidy up the spreadsheet after the meeting. It may be beneficial to have a scribe to assist you.
- Use the design architecture to break the FMEA work into pieces that would fit in one working session. Doing large FMEAs is tedious, and the process error prone, particularly if much time has passed in between sessions.
- It might be beneficial not to start the FMEA by displaying the complete FMEA template. Instead just focus on the items in the scope of analysis, their Failure Modes, mechanisms of failure and effects.
- Using a database of Failure Modes and End Effects is beneficial for speeding the work and creating consistency.

Other advices that could help with the efficiency of the FMEA process are as follows:

- Have a lead-person for the subject of analysis prefill the Failure Modes and mechanisms of failure as seeds for discussion during the FMEA working session. Caution should be taken in that sometimes seeing the answer to a problem could inhibit the creativity of the other team members from thinking of new Failure Modes and Causes.
- It may be useful to have the designers keep the FMEA template in mind as they do their design work. It could prompt them to think about Failure Modes while they are making design decisions.

Tip — Although FMEAs deal with failures under fault conditions, sometimes it may be beneficial to list an End Effect when there is no fault. See the two examples below:

Example 1: a sensor fails to detect low blood glucose. There is no fault—the sensor simply doesn't have 100% sensitivity.

Example 2: a polishing step leaves a pit in a metal surface. The process didn't fail. Normal metal surfaces sometimes pit.

In both examples, the End Effect is the same as if there was a failure. If there is no fault, there is also no Cause/mechanism of failure. In such cases write "No Fault" in the column "Causes/Mechanisms of Failure." This would be an extension of the FMEA for the benefit of risk management.

12.5 FMEA IN THE CONTEXT OF RISK MANAGEMENT

The hazard analysis process uses a confluence of FMEAs from lower system-levels into the RACT. The relation between FMEAs and the RACT is shown in the following figures.

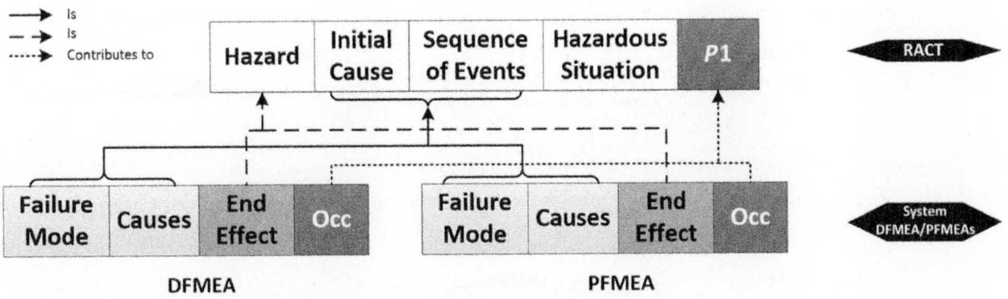

Figure 12.8 Integral Systems—System D/PFMEA to RACT Flow.

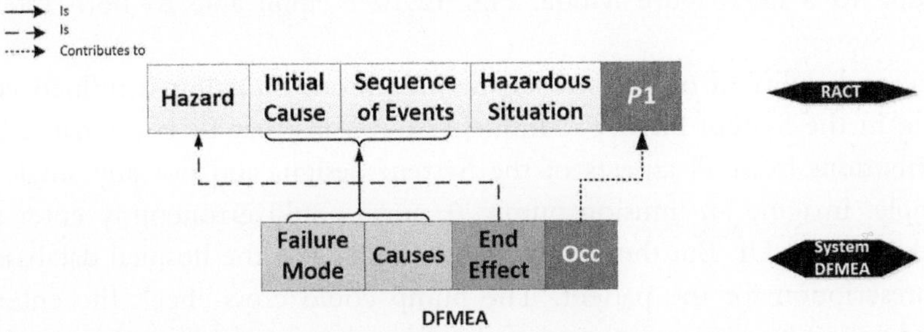

Figure 12.9 Distributed Systems—System DFMEA to RACT Flow

In Fig. 12.8 integral Systems are modeled (see Section 3.3 for the description of System types). For integral Systems, Hazards can come from product-design, or manufacturing-process failures. Failure Modes whose End Effects at the System level are Hazards are captured in the RACT as Hazards. Note that all System Hazards must be found in the CHL.

The initial Cause and sequence of events in the RACT are captured from the System FMEAs Causes, and Failure-Modes columns. Essentially, the Initial Cause and Sequence of Events tell the story of how a Hazard can be realized.

In Fig. 12.9 distributed Systems are modeled, where the relationship between the System DFMEA and the RACT is shown. The final assembly of distributed Systems is done by the user. Therefore there is no System-PFMEA. Errors by the user in the assembly of the System are captured in the UMFMEA (see Fig. 12.10). For distributed Systems PFMEAs are carried out up to Level 2, which are the highest integral components of the System (see Fig. 12.5 for a depiction leveling numbers).

A similar relationship exists between the UMFMEA and the RACT. Some of the End Effects of use-failures lead into Hazards. The End Effects which are Hazards, are captured in the RACT. Similarly, the initial Cause and sequence of events are captured from the UMFMEA Causes and Failure-Mode columns. The Initial Cause and Sequence of Events tell the story of how a Hazard can be

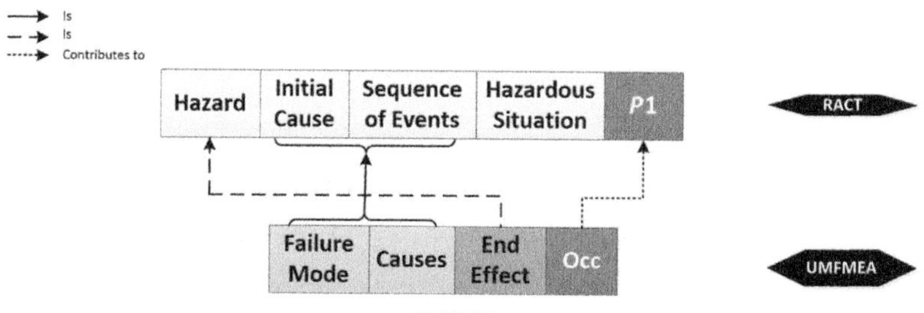

Figure 12.10 Relationship between Use-Misuse Failure Modes and Effects Analysis (UMFMEA) and the Risk Assessment and Control Table (RACT).

realized due to a use Failure Mode. Fig. 12.10 is applicable to both integral and distributed Systems.

P_1, the probability of occurrence of the Hazardous Situation is influenced by the Occ rating in the System FMEAs. "Influenced" means that P_1 is estimated based on the contributions from all aspects of the System design, and not any single FMEA. For example, imagine an infusion pump. A nurse could erroneously enter a wrong dose on the pump's UI. But the pump could be linked to the hospital database, which has the prescription for the patient. The pump could cross-check the entered dose against the patient's prescription, and if there is a large enough variation, the pump could alert the nurse to a possible error. In this scenario, Occ = the probability of the use-error, is not the same as P_1 the probability of patient exposure to an over/under dose. Also, P_1 is the probability of the occurrence of the Hazardous Situation. Occ rating from System FMEAs can only indicate the probability of occurrence of the Hazard, not the *exposure* to the Hazard(s).

12.6 DESIGN FAILURE MODES AND EFFECTS ANALYSIS

DFMEA is performed to assess the weaknesses that are present in the product and can be controlled by design. As with other FMEAs, DFMEA is a team activity. A suggested composition for the DFMEA team is:

- Design engineering
- System engineering
- Quality
- Clinical/Medical
- Risk management

It is best if DFMEA is done early in the design and development phase. Preliminary DFMEAs can and should be done on high-level designs, before all the details are known.

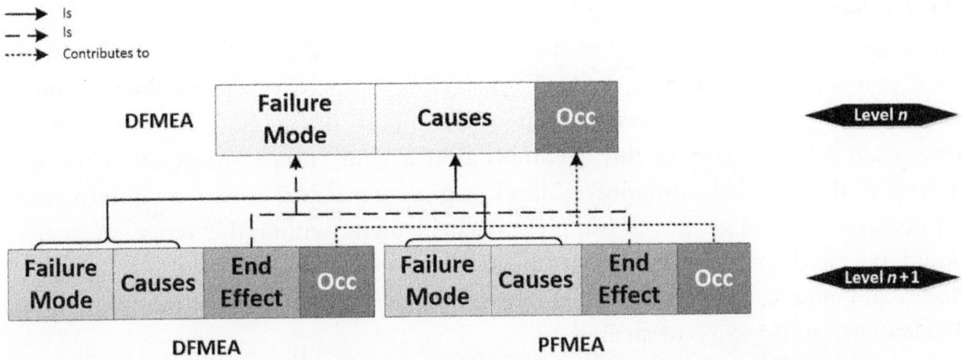

Figure 12.11 Information flow between Failure Modes and Effects Analysis (FMEA) levels.

Not every failure has a safety impact. FMEAs can be used for two benefits:

1. for *safety* risk analysis (Hazard Analysis), and
2. for *product* risk analysis. Product risks have impact on reliability, performance, and project, to name a few, but are distinguished from safety risks.

Each DFMEA/PFMEA at a given level contributes to the next level-up DFMEA. Failure Modes at a lower level DFMEA/PFMEA become Causes at the next level-up DFMEA. The End-effects at a lower level DFMEA/PFMEA become the Failure Modes at the next level-up DFMEA. Probability of Failure Modes at a lower level, contribute to the probability of a Failure Mode at the next level up (see Fig. 12.11 for a graphical depiction of this concept).

Per ground rule number 9 in Section 12.4.4, designer errors are excluded from the DFMEA. Excluding designer errors from the DFMEA does not mean that designers don't make mistakes. Designer errors are detected and corrected by *process*—which includes peer reviews, modeling, and testing.

It is possible to analyze the *design process* for ways in which designers could make mistakes and how those mistakes could escape. But that would be the PFMEA of the design process. The DFMEA analyzes the output of the design and the ways in which it could fail.

At the end of the design phase in product development, the DFMEAs should be transferred to Lifecycle Management and maintained by that department. Risk management should be kept apprised of any changes to the DFMEAs as part of any proposed change impact-analysis.

12.6.1 DFMEA workflow

In the following sections the workflow for DFMEA is described. The workflow corresponds to the template that is provided in Appendix B—Templates.

12.6.1.1 Set scope

The scope defines the boundary of the analysis, for the subject of analysis. For example, if you intend to analyze a defibrillator, include all parts of the product including packaging, in the scope of analysis. Then decompose the scope into its constituent elements. The granularity of this decomposition is your choice. You could choose to analyze at the detailed components level (e.g., a capacitor), or at an intermediate architectural level (e.g., power supply). Each element within the scope of analysis should be cited in the DFMEA template as an *item*, and analyzed for its Failure Modes and effects. It is permissible to have different depths of decomposition among the elements in the scope of analysis.

The scope of analysis should include the Failure Modes of each item, as well as Failure Modes of the interfaces among the items.

There are four types of interfaces between elements:

1. Physical interface—e.g., mechanical connections
2. Energy interface—e.g., electrical; thermal
3. Data interface—e.g., alerts; information; bit stream
4. Mass interface—e.g., fluids; gases

Energy, data, and mass are typically transmitted via physical interfaces among the elements, e.g., via wires, or tubes. Let's call the thing that is transmitted, "payload." For instance, the payload for a wire could be electrical energy, or data; and the payload for a tube could be a fluid or a gas. The physical interface should be considered an element in the scope of analysis. Let's examine the example in Fig. 12.12. System D is comprised of: elements A and B, pipe C and two connectors: 1, 2. Element A contains a fluid that is transmitted to Element B via pipe C. All three elements: A, B and C, and the connectors should be in the scope of analysis for System D.

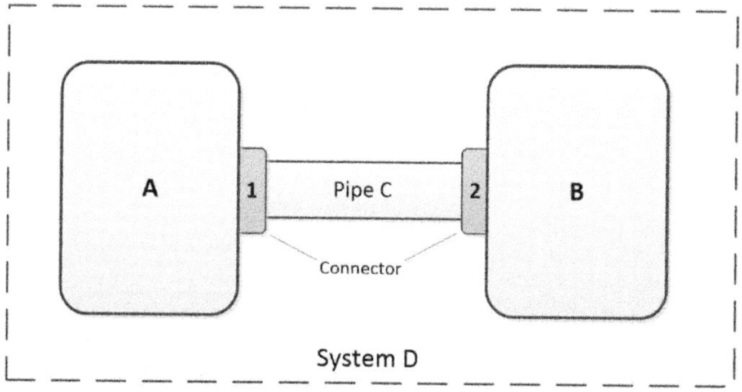

Figure 12.12 Interface example.

Now, let's assume we are performing a DFMEA of only element B. Element B requires the fluid for its function, else it fails. Lack of fluid could be due to the failure of element A to provide the fluid; a break in pipe C; or failure of the connectors. If we are using the hierarchical multilevel FMEA, B should be agnostic of the world outside of it. All B cares about, is that fluid is delivered to it. To B, lack of fluid is an external influence and would be cited as a Cause of failure. But B has no means of mitigating or controlling the supply of fluid.

There is one subtlety to consider. In whose scope of analysis should connector-2 reside? DFMEA of Pipe C, or DFMEA of element B? The decision is up to the analyst. A reasonable choice would be to include the part of the connection that is integral to B in the DFMEA of B, and the balance in the DFMEA of Pipe C.

Tip – There are certain components whose probability of failure is exceedingly small. For example, a properly designed wire that conveys a digital signal is not likely to fail while operating in its design environment. Therefore the contribution of failure of such a wire to safety risks would be negligible. In such cases, you can choose to exclude that element from the DFMEA.

12.6.1.2 Identify primary and secondary functions

The subject of analysis has a number of functions. Segregate the item's functions into primary and secondary subgroups. Primary functions are those that achieve the main mission of the subject of analysis. All other functions are secondary.

The reason for this action is that Severity ratings for the End Effect are influenced by the impact of the Failure Mode on the functionality of the subject of analysis.

12.6.1.3 Analyze

For each item in the scope of analysis identify its Failure Modes—answer the question: in what ways can this item fail to meet its design requirements? The Failure Modes could be functional or nonfunctional.

Example Failure Modes:

- Functional—doesn't perform its function, performs intermittently, late, early, too much, too little, etc.
- Nonfunctional—item swells, smokes, etc.

Consider the Failure Modes under normal use conditions, as well as reasonably foreseeable misuse conditions. For example, if a component is designed to operate in temperature range of 10–40°C, and it has been known that some users have operated it in temperatures of up to 50°C, then Failure Modes in the 10–50°C should be considered.

Each mode of failure of the item should go on a separate line in the template.

Identify the Causes/Mechanisms of Failure including the contributing initial Cause, and the chain of events that could lead to the Failure Mode. Include both internal Causes, such as aging of a part, as well as external Causes such as environmental temperature. Include failures in the interactions/interfaces among the elements within the scope of analysis.

Identify Local, and End Effects of the Failure Modes. An End Effect is that which is observable from outside of the boundary of analysis. A Local Effect is that which is not observable from outside of the boundary of analysis. It is possible for a Local-Effect to become the Cause for another Failure Mode. This is also referred to as Failure Mode interaction. In such cases, it is helpful to write a causal chain in the "Causes/Mechanisms of Failure" column, so it can be used again and built upon. Table 12.2 shows a snippet of a DFMEA where a Local Effect from row ID 1 becomes a Cause for the Failure Mode in row ID 2.

It is possible that a Failure Mode has only an End Effect, and no Local Effect. Denote this by entering N/A, none, or some notation to indicate the absence of a Local Effect. Leaving the cell blank could be misconstrued as incomplete analysis.

Tip – It is advisable to include in the cell for End Effects, any requirements that would be violated. For example, in Table 12.2, row ID 2, let's say there is a System requirement: Req123, that requires output pulses to be regular. Then cite *Req123* in the cell for End Effects. This is a convenience for the design team that would help them with the mitigation of the Failure Modes. And also, if a design change is proposed they can easily trace it back to the FMEAs.

Safety Impact is a System effect. To be able to determine whether a Failure Mode has a safety impact, we need to know how the subject of the analysis fits in the System. In the hierarchical multilevel FMEAs this can be known only after the integration of the FMEAs into the System DFMEA. But it may be possible to make some estimations of the Safety Impact in advance. For example, if it is certain that the Failure Mode would lead to one of the Hazards in the CHL, it would be a good guess that the Safety Impact will end up being Y. For instance, if the charging circuit in a defibrillator fails to charge the shock capacitor, likely the Safety Impact of that Failure Mode will be Y. Another way to estimate the Safety Impact of a Failure Mode is if it would violate a System requirement which is tagged as Safety.

If the Safety Impact of the Failure Mode cannot be determined in advance, you can set the Safety Impact to N as a generic setting and use the "No-Safety Impact" column in the Ratings tab of the template to determine the Severity rating. As the DFMEA is a living process and goes through an iterative process, when the FMEAs are rolled up to the System DFMEA, it will become apparent whether a given Failure Mode links up to any Hazards. After the integration of the FMEAs and creation of the System DFMEA, a cross-check is done to ensure consistency of Safety-Impact

Table 12.2 Sample design failure modes and effects analysis

	Item/Function		Potential failure modes and effects			
ID	**Item**	**Function**	**Failure mode**	**Causes/mechanisms of failure**	**Local effects of failure**	**End effects of failure**
1	Power Supply	Provide power to the device	Output voltage too low	High temp → C1 capacitor leaks	***CLK signal drifts***	Display dims
2	ASIC	Control stimulation	Irregular stimulation	High temp → C1 capacitor leaks → ***CLK signal drifts***	None	Irregular output pulses

ratings. Any End Effect that traces up to a Hazard must have a Y in the Safety Impact column.

Tip – Failure Modes whose Occurrence is indistinguishable from zero, are deemed not credible and thus are excluded from the DFMEA/PFMEA. However, to help the reviewers of the DFMEA/PFMEA know that the analyst has considered a particular Failure Mode, even though it is not credible, at the discretion of the analyst, it's permissible to cite such Failure Modes in the DFMEA. Failure Modes that are not credible should be clearly delineated and need not be further analyzed.

Assign ratings. Table 12.3, Table 12.4, and Table 12.5 offer suggestions for rankings of Severity, Occurrence, and Detectability, respectively. Let's examine each rating.

Severity is the significance of the worst reasonable consequence of the End Effect at the boundary of analysis. Severity ranking definitions are different depending on whether the End Effect has a safety impact or not. For End Effects that do not have a safety impact, use the left column in Table 12.3, and for those with a safety impact use the right column.

Table 12.3 Definitions of DFMEA severity ratings

	Severity Criteria (Sev)	
Rank	**Qualitative criteria—no safety impact**	**Qualitative criteria—safety impact**
5	*Catastrophic*: Described failure mode will cause immediate failure of the Subject. (Total loss of all functions—primary and secondary)	*Catastrophic*—Impact of the end-effect at the System level can be death
4	*Critical*: Described failure mode will severely impact Subject functionality | Complete loss of primary functions	*Critical*—Impact of the end-effect at the System level can be permanent impairment or life-threatening injury
3	*Serious*: Described failure mode will reduce Subject functionality. (Partial loss of primary functions | Complete loss of secondary functions)	*Serious*—Impact of the end-effect at the System level can be injury or impairment that requires professional medical intervention
2	*Minor*: Described failure mode will have temporal or self-restoring impact on functionality | partial loss of secondary functions	*Minor*—Impact of the end-effect at the System level can be temporary injury or impairment that does not require professional medical intervention
1	*None*: Described component failure will have no impact on functionality	*Negligible*—Impact of the end-effect at the System level can be at most an inconvenience, or temporary discomfort

To rank the severity of End Effects *without* a safety impact, it's necessary to know the primary and secondary functions of the item under analysis. For example, if a Failure Mode causes the complete loss of the primary functions, the severity ranking would be 4, Critical.

To rank the severity of an End Effect that *has* a safety impact, consider the effect at the System level. That is because to receive the benefit of a medical device, the users/patients interact with the System, not just the individual components of the System.

Table 12.4 Definitions of DFMEA Occurrence Ratings

Probability of Occurrence Criteria (Occ)

Category	Rank	Qualitative criteria	Quantitative criteria
Frequent	5	The occurrence is frequent. Failure may be almost certain or constant failure	$\geq 10^{-3}$
Probable	4	The occurrence is probable. Failure may be likely | repeated failures are expected	$<10^{-3}$ and $\geq 10^{-4}$
Occasional	3	The occurrence is occasional. Failures may occur at infrequent intervals	$<10^{-4}$ and $\geq 10^{-5}$
Remote	2	The occurrence is remote. Failures are seldom expected to occur	$<10^{-5}$ and $\geq 10^{-6}$
Improbable	1	The occurrence is improbable. The failure is not expected to occur	$<10^{-6}$

Table 12.5 DFMEA detectability ratings

Detection Criteria (Det)

Category	Rank	Qualitative criteria	Quantitative criteria
Undetectable	5	No detection opportunity | No means for detection | Countermeasures not possible	$<10^{-3}$
Low	4	Opportunity for detection is low | Countermeasures are unlikely	$<10^{-2}$ and $\geq 10^{-3}$
Moderate	3	Opportunity for detection is moderate | Countermeasures are probable	$<10^{-1}$ and $\geq 10^{-2}$
High	2	Opportunity for detection is high | Countermeasures are likely	$<9 \times 10^{-1}$ and $\geq 10^{-1}$
Almost Certain	1	Opportunity for detection is almost certain | Countermeasures are certain	$\geq 9 \text{ x } 10^{-1}$

As such, if the Failure Mode of the item under analysis could result in the System harming the patient, then the severity of that Harm is attributable to the End Effect of the Failure Mode in the DFMEA.

In the BXM method there are five classes of Harm severity. One could wonder how to choose the severity class of the Harm, when it may have a probability of Harm in all five severity classes. The answer is to choose the most probable severity class. Example: Let's say the failure of the sterile seal of an implantable device could cause contamination of the device, which could lead to infection. In the example that was presented in Fig. 12.13, Infection had the highest probability of causing a *Serious* Harm (72.1%). Therefore in the FMEA, the sterile seal failure would receive a severity ranking of 3, Serious. Remember that in the DFMEA, this ranking is used only for criticality determination and the prioritization of resources on mitigating Failure Modes. The risk of Hazards is computed in the RACT and is a different matter.

Tip – In the hierarchical multilevel FMEAs, lower level DFMEAs are associated with the physical components and are reusable. If a component is used in multiple Systems and is a contributor to Hazards in those Systems, it may be that the same component would have different severity rankings depending on the System in which it is used. The severity rankings in reused FMEAs will need to be adjusted in the context of the System in which they are reused.

Occ ratings can be estimated from a variety of sources. For example:

- Field failure data on the same product
- Failure data for similar items, used under similar conditions
- Published data, e.g., MIL-HDBK-338B
- Data from suppliers

If the quantitative criteria are used to estimate the Occ rating (see Table 12.4), ensure the probabilities have a meaningful basis for the subject of analysis. Examples: per-use; per device-year, etc.

In the context of risk management, detectability rating in DFMEA has a special meaning. It relates to *how likely it is for the End Effect to be detected and countermeasures be taken, external to the boundary of analysis, to minimize the risk of Harm.*

To elucidate this, consider the qualifier "external." Remember that per ground rule number 9 in Section 12.4.4, designer errors are excluded. It means that assuming the correct design of the subject of DFMEA, it may be that the End Effect could create a Hazard at the System level. How likely is it for the End Effect to be detected *and* a countermeasure be taken by an entity which is external to the subject of analysis, in order to minimize the risk of Harm?

Why "external?" If a Failure Mode is detectable *inside* the subject of the DFMEA, and the designer chooses to devise a mechanism to counteract that Failure Mode, then the new design *with* the counteracting mechanism becomes the subject of the DFMEA. This means internal detection is already built into the design. Example: a medical device is mains powered. If the leakage current exceeds a certain amount, the user will receive an electric shock. The excessive current leakage is detectable. The designer designs-in a circuit breaker that senses current leakage and cuts off the power to the medical device to prevent electric shock. The new design, including the circuit breaker, is now the target of analysis for the DFMEA.

The next point to notice is that "risk of Harm" is mentioned. Because we are using DFMEA at the service of safety, the focus is on the reduction of harm, not necessarily improvements on reliability, customer satisfaction, etc. For example, consider an electrosurgical device fails and delivers too much energy to the surgical site. If the End Effect of the Failure Mode is detectable by the surgeon, e.g., by an alarm in the device itself, or by observation of burning of patient tissue, then the surgeon can immediately disengage the device and apply medical care to the wound.

For Failure Modes that do not have a safety impact Detection is irrelevant from the risk management perspective. For such Failure Modes set the Det rating to 1.

Refer to Table 12.5 for definitions of detectability rankings. Use quantitative data if available. Otherwise use the qualitative criteria to determine the Detectability rankings.

RPN is a measure of criticality of a Failure Mode. RPN is the product of the rankings of Severity, Occurrence, and Detection. This number is used to prioritize the Failure Modes and determine the degree of compensation that must be exercised. Table 12.6 offers a suggested stratification of compensating actions based on the criticality of the Failure Mode. The boundaries in Table 12.6 are selected at 12 and 52. But it is up to the manufacturer to decide where to draw the boundaries. Table 12.6 says that for the highest segment of RPN ratings, Level 3, the RPN must be reduced.

For Level 2, RPN should be reduced as far as possible, for safety-related Failure Modes. But for nonsafety related Failure Modes, the decision as to how far to reduce the RPN is a business decision and depends on the feasibility of the actions needed to reduce the RPN.

Table 12.6 Design Failure Modes and Effects Analysis RPN table

RPN	Action
53–125	**Level 3**–Reduce RPN through failure compensating provisions.
13–52	**Level 2**–If Safety Impact is Y, reduce RPN to as low as possible. If Safety Impact is N, reduce RPN if feasible.
1–12	**Level 1**–If Safety Impact is Y, reduce RPN to as low as possible. If Safety Impact is N, further RPN reduction is not required.

For Level 1, per EN ISO 14971:2012 [7] for safety-related Failure Modes the RPN must be reduced as far as possible, therefore the treatment of RPN is the same as Level 2. However, for nonsafety related Failure Modes, further action is not required.

Reduction of criticality of Failure Modes is achieved via mitigations. Mitigations can eliminate the Failure Mode completely, reduce the likelihood of the Failure Mode, or diminish the severity of the End Effect of the Failure Mode. Examples of design means of mitigating Failure Modes are as follows:

- Use of redundancy, or backups
- Use of high-reliability parts
- Choice of proven, biocompatible materials

For the initial ratings, consider the design features that are already included in the design and serve to reduce criticality. Include those features in the Existing Mitigations column. If the initial criticality rating needs to be further reduced, suggest additional mitigations, and after they are implemented, reassess the criticality rating under the Final Rating group.

The Remarks column can be used to document rationales for the choices of ratings, or why further mitigations are not done for a Failure Mode with safety impact, or anything else that could help future reviewers of the DFMEA gain better understanding of the analysis.

In some exceptional cases, the *Failure Mode* and the *End Effect* could be the same. For an example see Fig. 12.13, which models a surgical robot (the System), that uses various Instruments. On some of the Instruments, a temperature sensor is mounted. The Instruments just carry the sensor and pass the sensor signal directly to the System to display the temperature inside the body. For each Instrument, the sensor becomes a component. Let's say a Failure Mode of the sensor is to fracture under force. When that happens, the sensor outputs the wrong voltage. This is reflected in the lower row in Fig. 12.13. The End Effect of the *Sensor FMEA* becomes the Failure Mode of the item: Sensor in the *Instrument FMEA* (*dotted arrow*). This is reflected on the middle row in Fig. 12.13. Now, the End Effect at the Instrument level is still the output of the wrong voltage. Continuing up to the System level, the wrong voltage causes inaccurate display of temperature. As you see in this example, the Failure Mode and End Effect of the sensor are the same in the context of the *Instrument FMEA*.

Although this construct may appear somewhat redundant and laborious—the citation of the sensor failure in the FMEA of the Instrument, creates the possibility of reusable FMEAs. A shortcut would have been to bypass the Instrument FMEA and cite the Sensor directly in the System FMEA. But this would defeat the reusability of the Instrument FMEA.

With the use of software tools for automation of FMEA work, adherence to the BXM method of hierarchical, modular, and reusable FMEAs becomes easier.

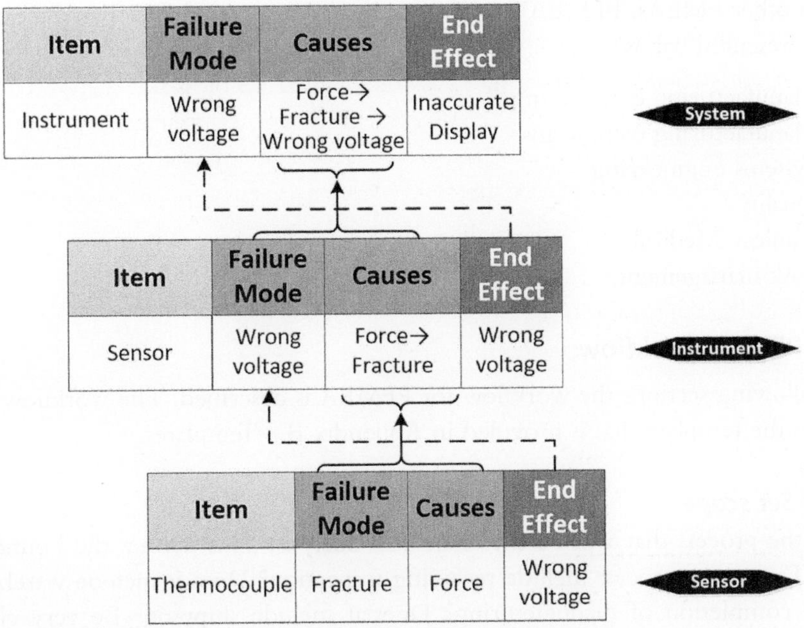

Figure 12.13 When End Effect and Failure Mode are the same.

12.7 PROCESS FAILURE MODES AND EFFECTS ANALYSIS

A process is a sequence of tasks that are organized to produce a product, or provide a service. PFMEA is a structured approach to identify weaknesses in process design, and assign criticality levels to each step in a process. PFMEA is a powerful prevention tool, since it does not wait for defects to occur, but rather anticipates them and implements countermeasures ahead of time.

For risk management, *Process* refers typically to manufacturing process, but could also include other processes such as service, repair, maintenance, and installation. For the balance of this section, the focus will be on manufacturing, but you can substitute any process that could have an impact on the safety of a medical device.

The PFMEA focuses on these questions:

- How can the process fail to deliver a product/part built to its specifications?
- What is the degree of criticality of a process step in producing a product/part that does not meet specifications?

Process Failure Modes with undesirable outcomes are mitigated via various means such as design or process changes. As a matter of practicality, the Failure Modes are prioritized so that the highest criticality Failure Modes are addressed first.

As in other FMEAs, PFMEA is a team activity. A suggested PFMEA team composition is presented below:

- Manufacturing engineering
- Manufacturing technicians
- Systems engineering
- Quality
- Clinical/Medical
- Risk management

12.7.1 PFMEA workflow

In the following sections the workflow for PFMEA is described. The workflow corresponds to the template that is provided in Appendix B—Templates.

12.7.1.1 Set scope

Identify the process that will be the subject of analysis. This defines the boundary of analysis. Does your process include receiving inspection? Does it include warehousing after the completion of manufacturing? Does it include shipping? Be very clear on what is included in the scope of analysis.

12.7.1.2 Identify primary and secondary functions

Identify the primary and secondary functions of the product of the process. Primary functions are those that achieve the main mission of the product of the process under analysis. In other words, primary functions are why the product is purchased. All other functions are secondary. For example, a pacemaker produces stimulation pulses to the heart, but also logs device faults. Both are functions, but producing pacing pulses is the primary function of the device, and fault logging is a secondary function.

The reason for this classification is that determination of the severity ratings of Failure Modes which do not have a safety impact, is based on the impact of the Failure Mode on the functionality of the product.

12.7.1.3 Process Flow Diagram

Process Flow Diagrams (PFDs) are a graphical way of describing a process, its constituent tasks, and their sequence. A PFD helps with the brainstorming and communication of the process design. The PFMEA process needs a complete list of tasks that comprise the process under analysis. The level of detail can be decided by the team. Including more detail takes time, but it reduces the probability of missing Failure Modes.

Manufacturing engineering should be able to produce the PFD. A good way to go about creating the PFD is to first log the major tasks of the process. Then add the detailed tasks, and the steps necessary to realize each task. Next, have a walk through the PFD with the stakeholders, e.g., manufacturing engineers and technicians, to debug the PFD. Perform the process using the PFD to verify the PFD. Add any discovered missing steps.

12.7.1.4 Analyze

Cite each step of the process that was described in the PFD, in the PFMEA template.

- Example process step: rinse casing.

For each process step, describe the purpose or intent of the step.

- Example purpose: to remove debris from machined casing.

For each process step, identify the ways in which it could fail. If there is more than one way to fail, enter each Failure Mode in a separate row.

Failure Modes of a process step could be:

- Complete failure
- Partial failure
- Intermittent failure
- Process drift, which could lead into failure

Process step failures can have one of four types of potential outcomes:

1. Desired outcome is achieved.
2. Desired outcome is not achieved.
   Example: rinse is incomplete.
3. Desired outcome is achieved but some deleterious unintended outcomes are also achieved. Example: rinse is complete, but rinse-head impacts casing.
4. Desired outcome is not achieved, and some deleterious unintended outcomes are achieved. Example: casing becomes contaminated due to use of wrong rinse solution.

Process step Failure Modes with outcomes 2, 3, or 4 are cited in the PFMEA.

In the "Causes/Mechanisms of Failure" column, describe *realistic* potential Causes of failure. Example: rinse timer drifts. Avoid highly imaginative but improbable Causes. While they may be interesting, or even entertaining, they don't add real value to the product development process and waste valuable engineering time in endless discussions.

The next two columns are End Effect and Local Effect. Each Failure Mode has an End Effect. An End Effect is that which is visible at the boundary of analysis. That is the impact of the Failure Mode which is perceivable on the product of the process. Example: metal debris on casing causes short circuit in the electronics. A Local Effect is not perceivable on the product of the process. It may be something internal that could cascade into another Failure Mode in a later process step, which would have its own End Effect. Example: rinse solution is not discarded. If the contaminated rinse solution is reused, it could be the Cause for another type of Failure Mode.

***Note*:** Limit yourself to the boundary of analysis, e.g., if the boundary of analysis is the process for manufacturing of a subassembly, evaluate the End Effect on the subassembly, not the assembly into which it goes.

Safety Impact is a System effect. To be able to determine whether a Failure Mode has a safety impact, we need to know how the product of the process fits in the System. In the hierarchical multilevel FMEAs, this can be known only after the integration of the FMEAs into the System DFMEA. But it may be possible to make some estimations of the Safety Impact in advance. For example, if it is certain that the Failure Mode would lead to one of the Hazards in the CHL, it would be a good guess that the Safety Impact will end up being Y. For example, if a toxic solvent is used as a process aid to create a part that will contact patient tissue; and the failure of a cleaning process step could leave toxic residues on the medical component, likely the Safety Impact of that Failure Mode will be Y. Another way to estimate the Safety Impact of a process-step failure is if it would violate a System requirement which is tagged as Safety.

If the Safety Impact of the Failure Mode cannot be determined in advance, you can set the Safety Impact to N as a generic setting and use the "No-Safety Impact" column in the Ratings tab of the template to determine the Severity rating. As the PFMEA is a living process and goes through an iterative process, when the FMEAs are rolled up to the System DFMEA, it will become apparent whether a given Failure Mode links up to any Hazards. After the integration of the FMEAs and creation of the System DFMEA, a cross-check is done to ensure consistency of Safety Impact ratings. Any End Effect that traces up to a Hazard must have a Y in the Safety Impact column.

Severity is the significance of the worst reasonable consequence of the End Effect at the boundary of analysis, and is ranked on two different scales: with a safety impact, and without a safety impact. Below, each scale is explained.

For nonsafety related Failure Modes, evaluate the severity at the boundary of analysis. That is, evaluate the impact of the Failure Mode on the product of the process under analysis. Use the column for "Nonsafety" in Table 12.7 to choose a ranking.

To rank the severity of an End Effect that *has* a safety impact, consider the effect at the System level. That is because to receive the benefit of a medical device, the users/patients interact with the System, not just the individual components of the System. As such, if the Failure Mode of the process step under analysis could result in the System harming the patient, then the severity of that Harm is attributable to the End Effect of the Failure Mode in the PFMEA.

Table 12.7 PFMEA severity rankings

Severity Criteria (Sev)		
Rank	**Severity descriptions—nonsafety**	**Severity description—safety impact**
5	Failure to meet Regulatory requirements | Total loss of all functions—primary and secondary | >70% of the production has to be scrapped	*Catastrophic*—Impact of the end-effect at the System level can be death
4	Loss or degradation of primary functions | Failure to meet product specification | Scrapping of 50%–70% of the production	*Critical*—Impact of the end-effect at the System level can be permanent impairment or life-threatening injury
3	Loss or degradation of secondary functions | Reduced reliability but still within Spec | Scrapping of 25%–50% of the production	*Serious*—Impact of the end-effect at the System level can be injury or impairment that requires professional medical intervention
2	Process Delay | Scrapping of 5%–25% of the production. | Minor cosmetic or usability impact but still within Spec	*Minor*—Impact of the end-effect at the System level can be temporary injury or impairment not requiring professional medical intervention
1	Scrapping of 0%–5% of the production. Some of the products have to be reworked	*Negligible*—Impact of the end-effect at the System level can be at most an inconvenience, or temporary discomfort

In the BXM method there are five classes of Harm severity. One could wonder how to choose the severity class of the Harm, when it may have a probability of Harm in all five severity classes. The answer is to choose the most probable severity class. Example: Let's say the Failure Mode of the process step that seals a sterile package for an implantable device, is "incomplete sealing." This could lead to contamination of the device, which could lead to infection. In the example that was presented in Fig. 12.13, Infection had the highest probability of causing a Serious Harm (72.1%). Therefore in the PFMEA, the Failure Mode of process step to seal the sterile package would receive a severity ranking of 3, Serious. Remember that in the PFMEA, this ranking is used only for criticality determination and the prioritization of resources on mitigating Failure Modes. The risk of Hazards is computed in the RACT and is a different matter.

The ranking in the Occ column of the PFMEA is indicative of the likelihood of occurrence of the Failure Mode. Refer to Table 12.8 for the definitions of the Occurrence rankings. Use quantitative data if available. Otherwise use the qualitative definitions to determine the Occ rank. If quantitative data is used, ensure the units are

Table 12.8 PFMEA Occurrence ratings

Probability of Occurrence Criteria (Occ)

Category	Rank	Qualitative criteria	Quantitative criteria
Frequent	5	The occurrence is frequent. Failure may be almost certain or constant failure	$\geq 10^{-1}$
Probable	4	The occurrence is probable. Failure may be likely | Repeated failures are expected	$<10^{-1}$ and $\geq 10^{-2}$
Occasional	3	The occurrence is occasional | Failures may occur at infrequent intervals	$<10^{-2}$ and $\geq 10^{-3}$
Remote	2	The occurrence is remote | Failures are seldom expected to occur	$<10^{-3}$ and $\geq 10^{-4}$
Improbable	1	The occurrence is improbable | Failure is not expected to occur	$<10^{-4}$

defined and consistently applied. The Occ rating is inclusive of the implementation of all pertinent mitigations. In other words, choose the Occurrence rank assuming the cited mitigations are already implemented and effective.

Det, or Detectability, is the likelihood of detection of a Failure Mode. It is an estimate of the probability of detecting the Failure Mode before the product is released. Therefore detection may occur anywhere in the causal chain, from the Cause of the Failure Mode to the End Effect. Refer to Table 12.9 for definitions of detectability rankings. Use quantitative data if available. Otherwise use the qualitative criteria to determine the Detectability rankings.

RPN is a measure of criticality of a Failure Mode. RPN is the product of the rankings of Severity, Occurrence, and Detection. This number is used to prioritize the Failure Modes and determine the degree of compensation that must be exercised. Table 12.10 offers a suggested stratification of compensating actions based on the criticality of the Failure Mode. The boundaries in Table 12.10 are selected at 12 and 52. But it is up to the manufacturer to decide where to draw the boundaries. Table 12.10 says that for the highest segment of RPN ratings, Level 3, the RPN must be reduced to a lower Level.

For Level 2, RPN should be reduced as far as possible, for safety-related Failure Modes. But for nonsafety-related Failure Modes, the decision as to how far to reduce the RPN is a business decision and depends on the feasibility of the actions needed to reduce the RPN.

For Level 1, per EN ISO 14971:2012 [7] for safety-related Failure Modes the RPN must be reduced as far as possible, therefore the treatment of RPN is the same as Level 2. However, for nonsafety-related Failure Modes, further action is not required.

Table 12.9 PFMEA detectability ratings

Detection Criteria (Det)

Category	Rank	Qualitative criteria	Quantitative criteria
Undetectable	5	Physics-of-Failure not understood | No detection opportunity (e.g. no inspection, or no means for detection) | Countermeasures not possible	$<10^{-3}$
Low	4	Failure is very difficult to detect | Opportunity for detection is low (e.g., very low sampling for inspection) | Countermeasures are unlikely	$<10^{-2}$ and $\geq 10^{-3}$
Moderate	3	Failure is moderately detectable | Opportunity for detection is moderate (e.g. 10% sampling) | Countermeasures are probable	$<10^{-1}$ and $\geq 10^{-2}$
High	2	Failure is very detectable | Opportunity for detection is high (e.g. 100% visual inspection) | Countermeasures are likely	$<9 \times 10^{-1}$ and $\geq 10^{-1}$
Almost Certain	1	Failure is obvious | Opportunity for detection is almost certain (e.g., 100% instrumented inspection) | Countermeasures are certain	$\geq 9 \times 10^{-1}$

Table 12.10 Process Failure Modes and Effects Analysis RPN table

RPN	Action
53–125	**Level 3**–Reduce RPN through failure compensating provisions.
13–52	**Level 2**–If Safety Impact is Y, reduce RPN to as low as possible. If Safety Impact is N, reduce RPN if feasible.
1–12	**Level 1**–If Safety Impact is Y, reduce RPN to as low as possible. If Safety Impact is N, further RPN reduction is not required.

12.8 USE/MISUSE FAILURE MODES AND EFFECTS ANALYSIS

UMFMEA is a tool to analyze the effects of failures in the use of a medical device, and also the effects of misuse of the device. We distinguish *Misuse* from *Use Failure.* As defined in Section 8.1, a Use Failure is the failure of a User to achieve the intended and expected outcome from the interaction with the medical device. A Use Failure is not something that the User intends, but Misuse is intended by the User.

The UMFMEA is a System-level FMEA, similar to System DFMEA or System PFMEA. This is because the User interacts with the whole System. Therefore the scope of analysis of the UMFMEA is the entire System, and End Effects of the UMFMEA can be System Hazards. The input to the UMFMEA is the set of System use scenarios, tasks, and step actions.

Just as in other FMEAs, the UMFMEA discovers many Failure Modes, only some of which may have a safety impact. Knowledge of the nonsafety-related Failure Modes is useful in improvement of the design for better user experience or product effectiveness. Risk management leverages only the Failure Modes that have a safety impact. The UMFMEA is an analytical tool that serves both the risk management and the usability engineering efforts.

Normally the UMFMEA considers the ways in which every task in the *normal flow* of events could go wrong. But users don't always follow the normal flow. Sometimes the users make mistakes and go down unexpected alternate paths. They may even improvise and create new pathways. Due to the fact that the number of alternate paths may be very large, it may be that task analysis doesn't consider all the possible alternate paths. It's advisable to try to consider the alternate paths that are related to safety critical operations of the device.

UMFMEA does not consider malice in the scope of analysis.

12.8.1 Distinctions

There are many special terms used in the domain of usability engineering and UMFMEA. It is important to have a clear understanding of these terms and their distinctions. Without this clarity, it would not be possible to properly analyze the medical device, or communicate your analysis. Below some of the important terms are examined.

Use: Using the device for what it was intended and per the supplied labeling.

The outcomes from attempted use can be:

- Successful use
- Failed use
  1. Step action is not performed.
     The user has the intention to perform action but is unable to complete the action.
     Example: UI does not permit the performance of the action, or UI is so confusing that the user cannot perform the action.
  2. Step action is performed, but with difficulty.
     The user has the intention, and executes the action but with difficulty.
     Example: Complicated UI causes the user to make mistakes that he/she recognizes and corrects the mistake. The action is ultimately completed but with struggle and errors.

3. Step action is performed incorrectly.
   The user has the intention and completes the task, but incorrectly. To analyze this outcome let's refer to Fig. 8.2. To isolate use-failures from design-failures, we'll focus on the right side of the picture. The potential causes of performing a step action incorrectly are:
   a. *Incorrect perception*—The User doesn't see/hear/feel the stimuli that is expected to be perceived, and therefore performs the action incorrectly.
   b. *Incorrect cognition*—The user has a wrong mental model of the system, and with the correct perception, performs the wrong action. Possible causes: lack of knowledge/training, unfamiliarity with the device. Or, user has a memory lapse. That is, the user knows what to do, but forgets, or makes a mistake.
   c. *Incorrect action*—The user has the correct perception, cognition, and intention but fails to perform the action correctly, e.g., a slip of the finger, or physical inability to reach/pull/twist/etc.

Failed use can be both due to action, or lack of action by the User. An unexpected outcome either in the patient physiology or in the medical device is not considered a Use Failure. Use the above three types of Failed-Use categories to guide in the distinction of Use Failure from other types of failures.

Abnormal use: Use of the device in a manner that goes beyond the Risk Controls which are put in place by the manufacturer. For example, alteration of a device, disregard of device alarms, or use of the device under conditions that are clearly prohibited by the manufacturer.

Example 1: Ventilator alarm system is intentionally disconnected, preventing detection of Hazardous Situation (Source: IEC 62366 [14], sec. C.3).
Example 2: Contrary to the indications in the accompanying documentation, a medical device is not sterilized prior to implantation (Source: IEC 62366 [14], sec. C.3).

Misuse: Incorrect or improper use of the medical device (Source: ISO 14971 [3,7]— unofficial definition).

Reasonably Foreseeable Misuse: Misuse is not use failure. It is deliberate and well intentioned. Abnormal or Misuse of the device that is *pervasive* and *well known* is Reasonably Foreseeable Misuse. Typically, off-label use of a medical device/drug for the benefit of patients is seen is foreseeable misuse.

Example: It has become routine to send patients home with intravenous (IV) infusion pumps. These pumps were originally designed for use in hospitals by trained medical professionals.

Malice: Use of a device with the intention to harm the patient or damage the device.

Table 12.11 Taxonomy of user actions
Use Scenario 1

Task 1	
	Step Action 1 Step Action 2 Step Action 3
Task 2	
	Step Action 1 Step Action 2

Use Scenario: Describes the interaction of a user with the device to achieve a certain outcome.

A Use-Scenario is comprised of one or more tasks. Each task is comprised of one or more action steps. Table 12.11 illustrates this taxonomy.

Tip — Use a verb or action phrase for Use-Scenario titles, Tasks, and Step Actions. Example: check status; deliver shock; remove pad from package.

12.8.2 Use Specification Versus Intended Use

ISO 14971 [3] defines *Intended Use* as the "intended purpose; use for which a product, process, or service is intended according to the specifications, instructions and information provided by the manufacturer."

IEC 62366 [14] defines *Use Specification* as

"summary of the important characteristics related to the context of use of the medical device

Note 1: the intended medical indication, patient population, part of the body or type of tissue interacted with, user profile, use environment, and operating principle are typical elements of the use specification.

Note 2: the summary of the medical device use specification is referred to by some authorities having jurisdiction as the "statement of intended use."

Note 3: to entry: the use specification is an input to determining the intended use of ISO 14971:2007."

While "Intended Use" and "Use Specification" are similar, they are not the same thing. *Intended Use* is akin to the clinical indication of the medical device—what disease; what part of the body, etc. *Use Specification* is about the context of use of the medical device, which includes:

- Intended use
- Intended user
- Intended use environment

12.8.3 UMFMEA Workflow

In the following sections the workflow for UMFMEA is described. The workflow corresponds to the template that is provided in Appendix B—Templates.

12.8.3.1 Set scope

Explicitly define the scope of analysis. For UMFMEA the scope should include the System and the users who would interact with the System, e.g., patient, physician, or service personnel. External influences to the scope of analysis are considered as causes to Failure Modes. For example, loud noises, dim ambient lighting, etc.

12.8.3.2 Identify primary and secondary functions

As in the DFMEA and PFMEA, identify the primary and secondary functions of the subject of analysis. Since UMFMEA analyzes the entire System, this step identifies the primary and secondary functions of the medical device. Primary functions are those that achieve the main mission of the System. All other functions are secondary. For example, if a pacemaker logs the history of therapy that it has delivered, the primary function of the pacemaker would be pacing, and the secondary function would be logging therapy history.

12.8.3.3 Analyze

Analysis of the System for use-failures requires knowledge of the System Use-Scenarios and their constituent tasks. A formal usability engineering task analysis is a good basis for the UMFMEA. If a formal task analysis is not available, at a minimum, inventory the Use Scenarios and actors. A graphical representation like Fig. 12.14 is a handy way to communicate and confer with your team to make sure a complete inventory of the use scenarios and actors is achieved.

Next, for each identified Use Scenario, list the tasks and step actions by the Actors. This can be tabular or a graphic flow chart. For each step action, hypothesize the ways in which the step action could be done incorrectly, and the causes thereof. Consider the distinctions that are described in Section 12.8.1. Use hypotheses such as: user is unable to see/hear, interpret, press, etc. Log this information in the UMFMEA spreadsheet.

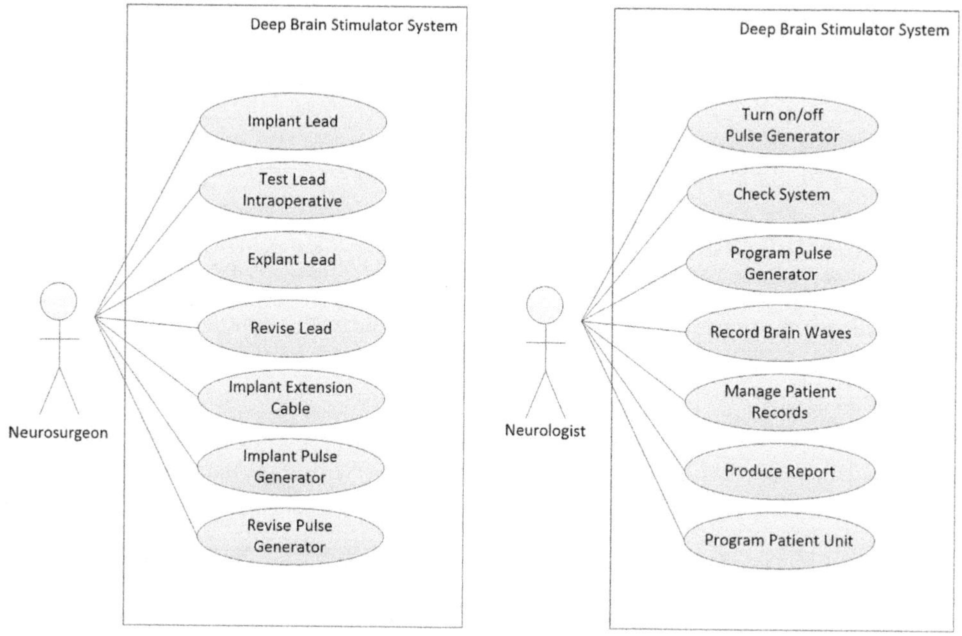

Figure 12.14 Use-Scenario Inventory.

List the potential effects of the Failure Mode. Local Effects may not be visible from outside the System, but create an internal effect within the System. End Effects are observable from outside the System.

If the End Effect is a Hazard in the CHL, then the Safety Impact is yes. Else, it is no.

List all the existing mitigation that serves to prevent the Failure Mode, in the "Existing Mitigations" column.

Assign ratings. Table 12.12, Table 12.13, and Table 12.14 offer suggestions for rankings of Severity, Occurrence, and Detectability, respectively. Remember that the ratings are inclusive of all the existing mitigations. Let's examine each rating.

Severity is the significance of the worst reasonable consequence of the End Effect at the System level. Severity ranking definitions are different depending on whether the End Effect has a safety impact or not. For End Effects that do not have a safety impact, use the left column in Table 12.12, and for those with a safety impact use the right column.

To rank the severity of End Effects *without a safety impact*, it's necessary to know the primary and secondary functions of the item under analysis. For example, if a Failure Mode causes the complete loss of the primary functions, the severity ranking would be 4, Critical. Use the information from Section 12.8.3.2 to determine the Severity ranking.

Table 12.12 Definitions of UMFMEA severity ratings

Severity Criteria (Sev)		
Rank	**Qualitative criteria—no safety impact**	**Qualitative criteria—safety impact**
5	*Catastrophic*: Described failure mode will cause immediate failure of the Subject. (Total loss of all functions—primary and secondary)	*Catastrophic*—Impact of the end-effect at the System level can be death
4	*Critical*: Described failure mode will severely impact Subject functionality | Complete loss of primary functions	*Critical*—Impact of the end-effect at the System level can be permanent impairment or life-threatening injury
3	*Serious*: Described failure mode will reduce Subject functionality. (Partial loss of primary functions | Complete loss of secondary functions)	*Serious*—Impact of the end-effect at the System level can be injury or impairment that requires professional medical intervention
2	*Minor*: Described failure mode will have temporal or self-restoring impact on functionality | Partial loss of secondary functions	*Minor*—Impact of the end-effect at the System level can be temporary injury or impairment that does not require professional medical intervention
1	*None*: Described failure mode will have no impact on functionality | Annoyance/inconvenience of the user	*Negligible*—Impact of the end-effect at the System level can be at most an inconvenience, or temporary discomfort

Table 12.13 Definitions of UMFMEA Occurrence ratings

Probability of Occurrence Criteria (Occ)		
Category	**Rank**	**Qualitative criteria**
Frequent	5	The occurrence is frequent. Experienced by almost every user
Probable	4	The occurrence is probable. Experienced by most users
Occasional	3	The occurrence is occasional. Experienced by some users
Remote	2	The occurrence is remote. Experienced by few users
Improbable	1	The occurrence is improbable. Has not been observed; not expected to be experienced by any user

To rank the Severity of an End Effect that *has* a safety impact, consider the impact on the patient/user. If you use a single-value severity ranking for Harms, then match that ranking to the right column of Table 12.12 and choose the ranking number.

In the BXM method there are five classes of Harm severity. As described in the DFMEA Section 12.6.1, choose the most probable severity class from the HAL. Example: Let's say the failure of the user to notice a damaged sterile seal of an

Table 12.14 Definitions of UMFMEA detectability ratings

Detection Criteria (Det)

Category	Rank	Qualitative criteria
Undetectable	5	Effect is not immediately visible or knowable | Countermeasures not possible
Low	4	Effect can be visible or knowable only with expert investigation using specialized equipment | Countermeasures are unlikely
Moderate	3	Effect can be visible or knowable with the moderate effort by user | Countermeasures are probable
High	2	Highly Detectable—Effect can be visible or knowable with simple action by user, from the information provided by the system itself | Countermeasures are likely
Almost Certain	1	Almost certain detection—Effect is clearly visible or knowable to user without any further action by user | Countermeasures are certain

implantable device could result in the implantation of a contaminated device, which could lead to infection. In the example HAL seen in Fig. 12.13, Infection has the highest probability of causing a Serious Harm (72.1%). Therefore in this example you would choose the severity ranking of 3, Serious. Remember that this ranking is used only for criticality determination and the prioritization of resources on mitigating Failure Modes. The safety risk of Hazards is computed in the RACT and is a different matter.

Occurrence ranking signifies the estimate of the highest likelihood of occurrence of a use failure. The Standard [14] indicates that quantitative estimation of Occurrence should not be made unless you have data to support your estimation. In the absence of data, you can make qualitative estimates of the potential for a use failure based on the descriptions in Table 12.13.

Det, or Detectability, is the likelihood of detection of the use failure. Detection may occur anywhere in the causal chain, from the initial use error, to an intermediate effect, to the final End Effect at the System level. Use Table 12.14 to select a ranking for Detectability. *Important: consider the ability of the user to take countermeasures to prevent the Hazardous Situation, or take actions to minimize Harm severity.* Detection of the End Effect where there is nothing that the user can do to minimize the risk of Harm is of little value. In such cases, choose a ranking of 5.

RPN is a measure of criticality of a Failure Mode. RPN is the product of the rankings of Severity, Occurrence, and Detection. This number is used to decide the

Table 12.15 Use-Misuse Failure Modes and Effects Analysis RPN table

RPN	Action
53–125	**Level 3**–Reduce RPN through failure compensating provisions.
13–52	**Level 2**–If Safety Impact is Y, reduce RPN to as low as possible. If Safety Impact is N, reduce RPN if feasible.
1–12	**Level 1**–If Safety Impact is Y, reduce RPN to as low as possible. If Safety Impact is N, further RPN reduction is not required.

degree of compensation that must be exercised. Table 12.15 offers a suggested stratification of compensating actions based on the criticality of the Failure Mode. The boundaries in Table 12.15 are selected at 12 and 52. But it is up to the manufacturer to decide where to draw the boundaries. Table 12.15 says that for the highest segment of RPN ratings, Level 3, the RPN must be reduced to a lower level. For Level 2, RPN should be reduced as far as possible for safety-related Failure Modes. For nonsafety-related Failure Modes, the decision as to how far to reduce the RPN is a business decision and depends on the feasibility of the actions needed to reduce the RPN. For Level 1, per EN ISO 14971:2012 [7] safety risks must be reduced as far as possible, therefore the treatment of RPN is the same as Level 2. However, for nonsafety related Failure Modes, further action is not required.

If it is decided to perform additional mitigations, list them in the "Additional Mitigations" column, and reestimate the ratings.

The Remarks column can be used to document rationales for the choices of ratings, or why further mitigations are not done for a Failure Mode with safety impact, or anything else that could help future reviewers of the UMFMEA gain better understanding of the analysis.

12.9 P-DIAGRAM

P-Diagrams, or Parameter Diagrams, are another tool that can be used at the service of Risk Management. P-Diagrams model a system and its behavior under various conditions. P-Diagrams can help with the development of FMEAs. The Error States can help with the identification of the Failure Modes in an FMEA, and Noise Factors can help identify the Causes of the Failure Modes in the FMEA.

Fig. 12.15 shows the construct of a P-Diagram. The main blocks in this diagram are the Input Signals, the System, the Control Factors, the Noise Factors, the Ideal Response, and the Error States.

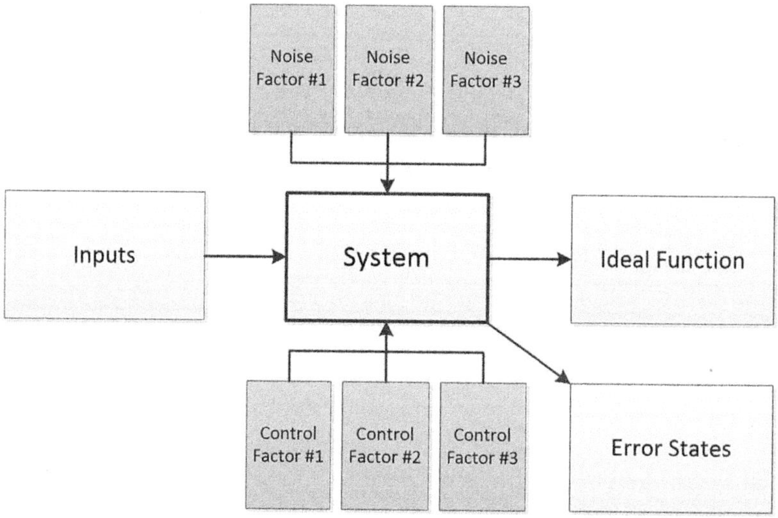

Figure 12.15 P-Diagram.

12.9.1 Input Signals

Input signals describe the items that the System needs to fulfill its objective. Examples of Input Signals: energy, data, and materials.

12.9.2 System

The System is the entity that processes the Inputs, under the Control Factors and the Noise Factors, and delivers the output.

12.9.3 Control Factors

Control Factors are the factors over which we have control, and can be changed as desired. Examples: quantity of input materials, temperature of oven, duration of curing, etc.

12.9.4 Noise Factors

Noise Factors are things that can influence the output of the System, but over which we do not have control. Examples: piece-to-piece variation, unintended customer usage, environmental conditions.

Noise Factors are typically grouped in five categories:

- Piece-to-piece variation
- Change over time/mileage
- Customer usage/duty cycle
- External Environment
- System interaction

12.9.5 Ideal Function

This is the intended output of the System—the way the designers designed the System. Naturally, things don't always work the way you want them to, as Noise Factors play a role.

12.9.6 Error States

These are the unintended outputs of the System. They are the result of Noise Factors' influence on the operation of the System.

12.9.7 Workflow

Use the following steps to create a P-Diagram

1. Identify the intended function of the System under analysis, and the expected Outputs
2. Identify the Input Signals—what are the inputs to the System? The things that the System needs in order to produce its output—the Ideal Function.
3. Identify the Error States—what are the ways in which the System can produce an output which is different from what is expected?
4. Identify the Noise Factors—what are the inputs, or influencers to the System function, over which you don't have control?
5. Identify the Control Factors—what are the inputs, or influencers to the System function, over which you do have control?

P-Diagram analysis provides a systematic way to consider a function's Error States and Noise Factors, as well as the methods that can be used to control them.

As in FMEA, the level of granularity of analysis is the analyst's choice. At the highest level "System" would be the entire medical device system. Or, "System" could be subsystems, or lower level components of the System.

P-diagram analysis can help you identify Hazards and their Causes. By examining the Noise Factors, you can create a causal chain that would explain how a hazardous output could be realized. You will likely find overlaps between FMEA and FTA findings and P-Diagram analysis. Ultimately it is your choice as to how many tools to use. While extra analyses consume more resources, they also reduce the likelihood of missing some Hazards and their causal chains.

12.10 COMPARISON OF FTA, FMEA

FTA and FMEA are both useful and important analytical tools in risk management. They both play a role, and it is best if they are used in a complementary fashion. As each tool has strengths and weaknesses, the combinatorial application leverages the best that each tool has to offer.

FTA is a top-down deductive analysis tool that starts with the top event, e.g., a Hazardous Situation, and works backwards toward the root cause(s). It seeks to answer the question: "How can this Hazardous Situation occur?"

FMEA is a bottom-up inductive analysis tool that starts at the basic elemental level and works forward toward the top event to answer the question: "what is the End Effect of the failure of the item in question?"

The FTA is more suitable for:

- Early in the product development process, when only high-level knowledge of the device is available.
- When there are few top events of interest, e.g., for derivative products when the predicate device is already well analyzed and understood, and the derivative only adds a few new top events.
- When a top event can be caused by multiple initiating Causes, or where there are many interactions and relationships among the components.
- Detection of CCFs.
- Systems with redundancies in the design.

The FMEA is more suitable for:

- Systems that are novel, or complex, and not well understood.
- When there are a large number of top events that can result from bottom events.
- Where occurrence of the top events do not require multiple faults.
- When there is a need for fail-safe operation.

First-order cut sets in the FTA, should also appear in FMEAs as single-point failures that can result in the top event.

Tip — Use FTA to prioritize FMEA work for complex products. FTA can more efficiently identify safety-critical parts of the System. Use this knowledge to prioritize the FMEA work.

CHAPTER 13

Safety Versus Reliability

Abstract

There is a general misperception that reliability and safety are the same, and that a reliable device is a safe device. While this is sometimes true, it is not a universal truth.

Keywords: Safety; reliability

There is a general misperception that reliability and safety are the same, and that a reliable device is a safe device. While this is sometimes true, it is not a universal truth.

A simple example is a scalpel. A reliable scalpel will cut tissue every time, but if it cuts the surgeon's hand, there is a safety issue. A converse example—imagine a medical device that has so many safety mechanisms that it frequently shuts down to avoid causing injury. The device may be very safe, but also very unreliable. Typically, when safety mechanisms have high sensitivity but low specificity, we encounter safe but unreliable Systems.

The role that reliability plays in safety is when both attributes are aligned. Take the example of a stent. An unreliable stent that dislodges, or fractures, would be also an unsafe stent. When safety and reliability are aligned, improved reliability also improves safety.

Safety Risk Management for Medical Devices
DOI: https://doi.org/10.1016/B978-0-12-813098-8.00013-1

© 2018 Elsevier Ltd.
All rights reserved.

CHAPTER 14

Influence of Security on Safety

Abstract

Medical devices are becoming progressively more connected to other healthcare systems, and the cybersecurity attack surface is becoming larger. Security exploits can be used to harm patients in many ways. Safety impact of cybersecurity exploits must be considered in the overall residual safety risk of medical devices.

Keywords: Cybersecurity; exploits; safety; security threat

Medical devices are becoming progressively more connected to other healthcare systems, and the cybersecurity attack surface is becoming larger. Security exploits can be used to harm patients in many ways. Because this book is focused on the safety risks of medical devices, we will stay focused on the impact of security on the safety risks of medical devices.

Security risk analysis is a parallel and related process to safety risk management. With respect to safety, security is another potential source of Hazards. The types of security-related Hazards include:

- Unavailability—a security attack may cause the medical device to become unavailable. In some cases, e.g., life-supporting medical devices, this would create a Hazard.
- Change of programming or code—a security attack could alter the code or programming parameters of a medical device, thus altering its behavior or performance.

As seen in Fig. 14.1, there is an inverse direction of impact when safety and security aspects are compared. In the domain of safety, humans are potentially harmed by Hazards from the medical devices. For security, the medical device is the potential target of harm by intended or accidental attacks carried out by humans. The security of a medical device might impact its safety, and loop back to harm the humans.

Security threats can be divided into three groups:

Group 1. Intentional—Malicious (aims to harm)
Group 2. Intentional—Misuse (aims to do good; e.g., get around a cumbersome user interface)
Group 3. Unintentional—Use error

Safety Risk Management for Medical Devices
DOI: https://doi.org/10.1016/B978-0-12-813098-8.00014-3

© 2018 Elsevier Ltd.
All rights reserved.

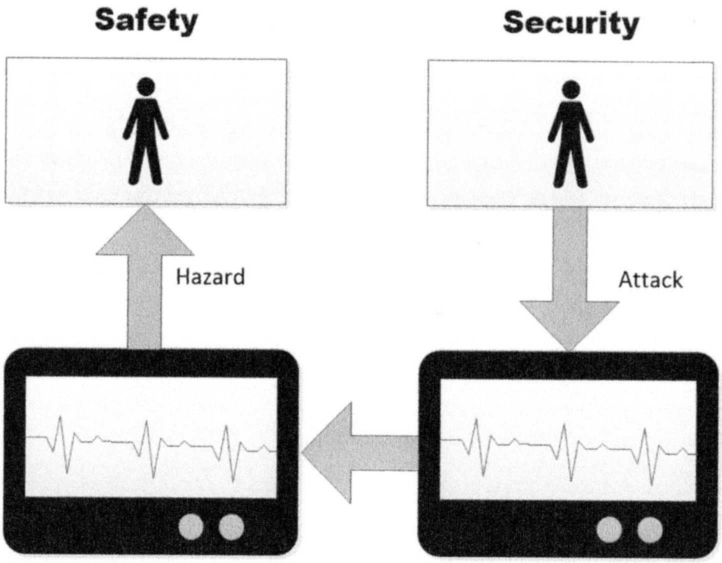

Figure 14.1 Safety and security relationship.

Traditionally, malicious actions are excluded from safety risk analysis, i.e., the hazard of someone using a medical device as a weapon is not included in safety risk analysis. However, because malice is a normal and expected part of security risk threats, we will include Group 1 in the safety risk impact assessment.

Group 2 is a valid input from Security Risk Analysis that should be captured in the Causes/Mechanisms of Failure in the Use-Misuse Failure Modes and Effects Analysis (UMFMEA). In this case, the security threat can potentially damage a function of the System, which may then have a safety impact.

Group 3 is already covered under UMFMEA.

Risk management involves the identification of Hazards and the estimation of risks due to those Hazards. Risk estimation requires knowledge of the probability of occurrence of Hazardous Situations. In the case of security-related Hazards, estimation of probability of an exploit is very difficult. Conventional wisdom suggests using motivation as an indicator of the likelihood of an attack. But experience has shown that motivation plays a smaller role than people think. For many attackers, the challenge of a break-in and the thrill of the exploit is enough reward.

It's important to consider that while security threats can have an adverse safety impact, the security *Risk Controls* themselves could create safety Hazards. Estimating the probability of occurrence of a Hazard from a security Risk Control is as difficult as estimating the probability of occurrence of a software failure.

The FDA has released a guidance titled: Postmarket Management of Cybersecurity in Medical Devices [25]. In this guidance, the FDA states that "estimating the probability of a cybersecurity exploit is very difficult due to factors such as; complexity of exploitation, availability of exploits, and exploit toolkits." The guidance [25] suggests, that in the absence of probability data, to use a "reasonable worst-case estimate" and set the value of probability of occurrence of the Hazardous Situation to 1. Alternatively, the FDA suggests that manufacturers instead use a "cybersecurity vulnerability assessment tool or similar scoring system for rating vulnerabilities and determining the need for and urgency of the response."

To handle safety risks from security exploits, estimate the vulnerabilities of the medical device to security threats, using means such as a cybersecurity vulnerability assessment tool. With that knowledge, assuming the exploit has happened, estimate the worst-case Hazard from the exploit. Further, assume the probability of exposure to the Hazard is 100%, and identify the potential Harms. To determine the acceptability of security-related safety risks, in the absence of quantifiable data, Ref. [25] suggests using a qualitative matrix that combines exploitability versus Harm severity. Fig. 14.2 is from Ref. [25], which can be used as a model. It indicates a fuzzy boundary between controlled and uncontrolled risks. The specific construct of the matrix for different applications is left up to the manufacturer.

If you use a single value for Harm severities, plot the security threat's exploitability versus Harm severity in the matrix. In the BXM method, five probability values are given for each Harm in the Harms Assessment List, one for each severity class. Choose the severity with the highest likelihood, and similarly plot the security threat's exploitability versus Harm severities.

Triage the security threats based on the matrix, from the most critical to the least critical, and address them in the order of urgency using the best available methods and tools. This way the security-related safety risks are reduced to as low as possible.

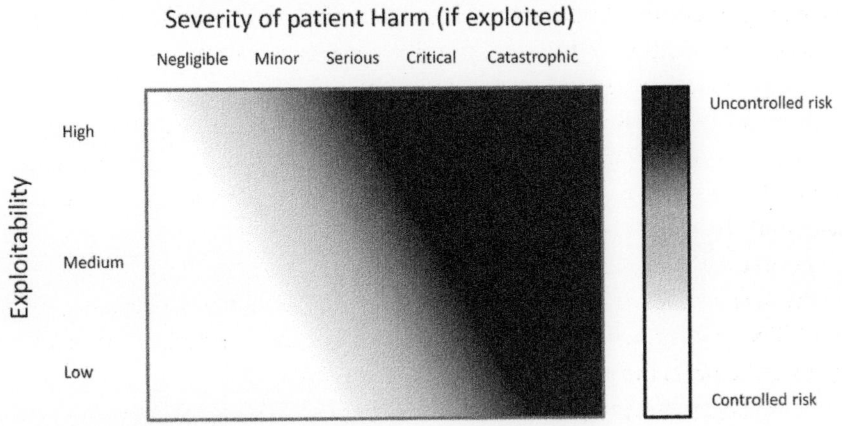

Figure 14.2 Exploitability versus Harm severity.

Next, analyze the safety risk potential from the security Risk Controls themselves. Reduce the safety risks due to the security Risk Controls to as low as possible.

While it is difficult to predict the probability of a cybersecurity exploit, it may be possible to estimate the probability of Hazards due to the *security Risk Controls*, because they are implemented under the manufacturer's control. For example, if an encryption algorithm is used to protect certain data, it may be possible to estimate the probability of an error in the encryption/decryption process.

By this point, safety risks due to security threats are reduced to as low as possible. If the overall risks of the product are found to be acceptable, and the product is released, postmarket risk management processes should be used to maintain and update the risk management file with respect to security threats in the same manner as other safety risks.

CHAPTER 15

Software Risk Management

Abstract

Software can have a strong influence on the safety of medical devices. This includes new software as well as legacy software and SOUP. IEC 62304 offers guidance and strategies that support the creation of safer software. These strategies in concert with ISO 14971 allow management of risks due to software failures. SFMEA as one of the tools of software risk management is expounded in this chapter, and special tips are offered for successful development of safety-critical medical software.

Keywords: Software risk management; Software FMEA; SFMEA; legacy software; SOUP; software safety classification

Software can have a strong influence on the safety of medical devices. Software for complex Systems is difficult to correctly specify, implement, and verify. Errors in software requirements specification, and software design and implementation are the main contributors of software-caused System Hazards. The most effective way to manage risks due to software is to consider the role of software before System design is completed.

Though deterministic, software is not necessarily predictable for complex Systems. This makes risk management of software a particularly difficult challenge. IEC 62304 [9] suggests three means to manage software risks:

1. Safety Classification, Risk Controls, and rigor in software development
2. Software configuration management
3. Software problem resolution process

Safety classification of software is covered in Section 15.3. Software configuration management and problem resolution process are/should be part of the Quality system, and are not discussed in this book.

With respect to functionality in medical devices, three types of software can be identified:

1. Software that provides clinical function
2. Software that is used as a Risk Control measure
   a. For hardware failures
   b. For software failures
   c. For use-failures
3. Other software
   a. Software whose failure could create a safety impact
   b. Software whose failure would *not* create a safety impact

DOI: https://doi.org/10.1016/B978-0-12-813098-8.00015-5

© 2018 Elsevier Ltd.
All rights reserved.

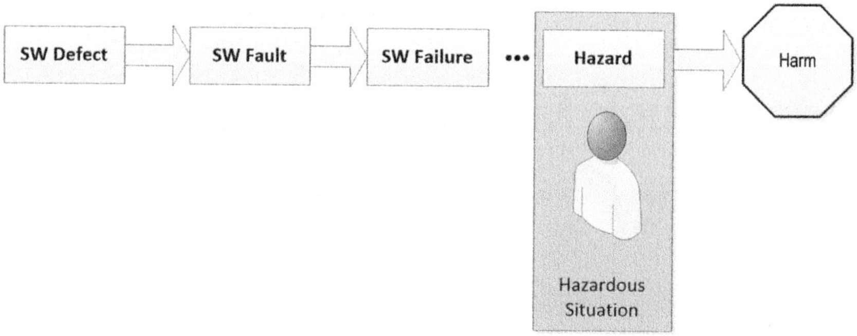

Figure 15.1 Contribution of software to Hazards.

Of the above types of software only 3.b is not safety related, and the rest should be included in the software risk management process.

As depicted in Fig. 15.1, Hazards are the consequence of a chain of events. Exposure to Hazards can cause Harm. Since exposure to software cannot cause Harm, software itself is not a Hazard. But software failures can cause Hazards in the System context. To determine the risks due to software, we'll need to identify the Harms that could result from software failures.

Understanding the contribution of software to Hazards and Harms can be achieved via top-down System analyses, e.g., Fault Tree Analysis. This requires knowledge of the System architecture, indication for use, and the context within which the System operates. Starting with potential Harms of the System deductively analyze the pathways which could lead to those Harms. If no pathway contains software, you can conclude that software in that System is not a contributor to Hazards.

If it is established that software is a contributor to Hazards and device risks, it is beneficial to analyze the software architecture to determine the roles and contributions of the various software items to risk. For simple devices, the manufacturer may choose to treat the software as a black box and bypass the software architectural analysis.

Before we proceed, it's important to understand some relevant vocabulary. Three important terms are:

Software System	An integrated collection of Software Items organized to accomplish a specific function or set of functions [9]
Software Unit	Software Item that is not subdivided into other items [9]
Software Item	Any identifiable part of a computer program, i.e., source code, object code, control code, control data, or a collection of these items [9] All levels of software composition can be called software Item, including the top-level: Software System, and the bottom-level: Software Unit

The analysis of software at the software architectural level enables us to plan architectural-level Risk Controls, such as introducing protective Risk Control measures that are external to the software. Beyond introduction of architectural Risk Control measures, applying the methods that are prescribed in IEC 62304 [9] provides process-based Risk Control measures that can reduce the probability of software failures.

IEC 62304 [9] B.7 says "Software risk management is a part of the overall medical device risk management and cannot be adequately addressed in isolation." Since software is not a Hazard by itself, it doesn't have risk. Risks of software failures can only be estimated in the context of the System. *Software risk in a System is the aggregation of the risks of all the Hazards that are caused by software failures.*

Software safety must be approached within a multidisciplinary context including: System design, software engineering, mechanical and electrical design, and usability engineering.

Before delving deeper into software risk management, certain key terms need to be defined.

Software Defect: An error in design/implementation of the software.

Software Fault: A software condition that causes the *software* not to perform as intended.

Software Failure: A software condition that causes the *System* not to perform according to its specification (System here means the medical device).

The following clarifications should facilitate a deeper understanding of the above terms:

- A software defect does not necessarily cause a software fault
- A software fault does not necessarily cause a software failure
- A software failure does not necessarily create a System Hazard
- Software could fault in the absence of software defects (e.g., due to bad software requirements)
- Software could fail in the absence of software faults (e.g., due to bad System requirements)

Software defects are systemic, not random in nature. Software testing can catch a proportion of software defects. But software of even moderate complexity cannot be completely tested. Therefore we have to accept the fact that some software defects will probably remain in the software.

Fig. 15.2 shows the chain of software events that could lead into a System Hazard. The *dotted arrows* imply that there may be intervening events between software failure and System Hazards.

In the context of *software risk management*, in this book we will only consider risks due to software failures. In other words, the progression depicted by the *dark arrows* in

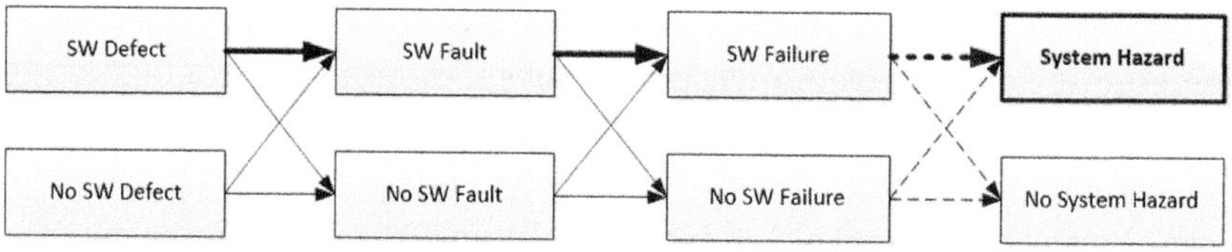

Figure 15.2 Software chain of events to system hazards.

Fig. 15.2 is considered because that is where software risk management can have an influence. The *light arrows* indicate either no problems with software, or no System Hazards, and are thus not considered in software risk management. As shown in Fig. 15.2, even when the software works perfectly according to its specification, the System can present safety risks. Those risks are managed in the *System risk management* per ISO 14971 [3,7].

Risk management per ISO 14971 [3,7] requires planning. Likewise, software risk management requires planning and documentation. The software risk management planning documents can be separate or combined with the System risk management documents.

As in other kinds of risk management, management of software risks entails the identification of Hazards, estimation of risks of the Hazards, control of the risks, and evaluation of the risks. In the sections below elements of software risk management such as software safety classification, software hazard identification, and topics unique to software such as handling of Software of Unknown Provenance (SOUP) are discussed.

15.1 SOFTWARE RISK ANALYSIS

Software risk analysis starts with the identification of the intended use of the System. Without this knowledge, it's not possible to determine whether a particular software failure is hazardous or not. Knowing the intended use of the System and the System architecture, it is possible to postulate potential Hazardous Situations and analyze for contributions of the various elements of the Systems, including software. This activity is not truly completed until the software architecture is determined and software items are identified. With the knowledge of software architecture, it becomes possible to determine the contribution of individual software items to Hazardous Situations.

Analysis of risk involves the identification of Hazards, and estimation of their risks. As in hardware risk-analysis, in software risk-analysis, Hazards must be identified and their risks are estimated. In hardware, Hazards are sometimes rooted in physical hardware failures. But software doesn't fail in the same ways as hardware. It doesn't wear out, fatigue, or just break.

Table B.1 of IEC TR80002-1 [26] offers examples of Causes of software faults that could introduce Hazards. The Causes are grouped by functional areas and guiding questions are provided to help the analyst uncover missing or inadequate safety requirements. Table B.2 of Ref. [26] identifies examples of software failure Causes that can have broad impact on the System operation. Examples: divide by zero or errant pointers. With these kinds of errors, failures that originate in nonsafety-critical software items can impact safety-critical software items. Table B.2 [26] offers suggestions for verification methods to trap such failure Causes. For these types of software failures, requirements-based testing is not very effective.

An analytical tool that can be used to identify software-related Hazards is the Software Failure Modes and Effects Analysis (SFMEA). Section 15.2 describes the usage of this tool.

Estimating the risk of a Hazard involves estimating the likelihood of occurrence of the Hazard. In the case of software-caused Hazards, there is no consensus on a method to estimate the probability of occurrence of software failures. The Standard [9] advocates for the identification of Hazards that could potentially be caused by software, and implementing Risk Control measures, instead of trying to estimate the risk of software failures.

To get some sense of the relative risks from software-caused Hazards, the Standard [9] suggests relying on the severity of Harms alone. However, IEC 62304 [9] acknowledges that it may be possible to quantitatively estimate the probability of occurrence of software failures for *legacy* software, based on the usage of legacy software and examination of postproduction data.

Medical device software could be classified into two categories: new software or legacy software. Legacy software is software that has been in use in the field and for which postproduction history data may be available. New software is software that has not been released for use in the field yet and has no postproduction history. This creates two distinct pathways for software risk management:

Case 1. where the probability of software failure can be estimated—legacy software

Case 2. where the probability of software failure cannot be estimated—new software, or legacy software for which postproduction data is unavailable, or is of questionable quality

In Case 1, software risk management process is very similar to hardware risk management. In Case 2, however, the process for software risk management is different because without knowledge of the probability of occurrence of software failure, it is not possible to estimate the risk of a software-induced Hazard. In such cases, the Standard [3,7] recommends to assume P(software failure) = 1 and instead focus on the rigors of design and development process, and implementation of Risk Controls.

The Standard IEC 62304 [9] recognizes that application of appropriate levels of rigor to software development does reduce the probability of failure of software items, presumably due to detection and elimination of software defects.

It should be understood that setting P(software failure) = 1 doesn't necessarily mean that $P_1 = 100\%$. It means: if software is an element in the causal chain that leads to the exposure to a Hazard, set the probability of software failure to 100%.

$$\text{Risk} = P(\text{Hazardous Situation}) \times P(\text{Harm}) = P_1 \times P_2$$
$$P_1 = P(\text{Hazardous Situation}) = P(\text{Hazard}) \times P(\text{Exposure})$$
$$P(\text{Hazard}) = P(\text{Software Failure}) \times P(\text{additional intervening events})$$

Using the above equations, *risk* of software failures can be computed. If *P*(Software Failure) is set to 1, then *P*(Hazard) = *P*(additional intervening events).

In some Systems a software failure may immediately expose the patient/user to a Hazard. In such cases, *P*(Software Failure) = *P*(Hazardous Situation).

If *P*(Software Failure) is unknown, the risk of Harm due to software failure cannot be estimated and thus cannot be included in the overall residual risk computations.

15.2 SOFTWARE FAILURE MODES AND EFFECTS ANALYSIS (SFMEA)

Failure Modes and Effects Analyses (FMEAs) are a common and ubiquitous tool for hazard analysis. SFMEA is a variation of the Design Failure Modes and Effects Analysis (DFMEA). When used for software hazard analysis, FMEAs are applied in a slightly different manner than in hardware FMEAs. SFMEAs are applied to software architectural elements, or software items. This requires the knowledge of the software architecture and inputs to the software.

Systemic Causes/mechanisms of software failure such as design or implementation errors, or hardware anomalies, like bit flips are analogous to common cause failures in hardware and should be mitigated globally for the whole software system, not cited for every row of the SFMEA.

15.2.1 Software Failure Modes and Effects Analysis Workflow

The following explanation of the SFMEA is based on the SFMEA template that is provided in Appendix B—Templates.

Entries in the "Item" column are the elements within the scope of analysis. "Source" column captures where the Failure Mode was identified. In hierarchical multilevel FMEAs, this refers the underlying FMEAs whose End Effects were rolled up to the current SFMEA.

The entries in the "Failure Mode" column of the FMEA would be the answers to the question: "In what ways can this software item fail to perform its intended function?" One way to do this is to list the requirements of the software item and consider how not meeting each requirement affects the software item's intended function.

Software Failure Modes can have direct and/or indirect Causes. Direct Causes are local to the software item at hand, and do not necessarily affect other software items. Examples: incorrectly implemented algorithms, errors in input processing, and errors in output processing of the software item. Indirect Causes include: stack overflow, uninitialized pointers, and race conditions. Indirect Causes are more unpredictable than direct Causes. Other indirect Causes of software failures are: bit flips, low-power condition, or software sources such as operating systems, libraries, and SOUP.

Unlike DFMEA, SFMEA entries in the column "Causes/Mechanisms of Failure" would not be things like aging, fatigue, or wear out. Items that go in the column "Causes/Mechanisms of Failure" are external factors or systemic Causes. Since the object of the analysis is a software item, the question would be: "what factors could cause this software item not to perform its function?" For example, consider a software item that is supposed to pressurize a tank to 10 psi and a pressure sensor provides the tank pressure as input to the software item. If the pressure sensor input to the software item is incorrect for any reason, the software item would fail to meet its function.

The entries in the "End Effect" columns would be the consequences of the software item's Failure Mode at the boundary of analysis. Boundary of analysis is chosen by the analyst, and could be, e.g., a software subsystem, the entire software system, or the medical device. If the scope of analysis is the software system (as a component of the medical device), then the question is: "How would the Failure Mode of the software item in question affect the behavior of the software system?" If on the other hand the scope of analysis is the medical device, then the question would be: "How would the Failure Mode of the software item in question affect the behavior of the medical device?" Local Effects are not visible from outside the boundary of analysis but are noteworthy because they may cascade into other End Effects. It is possible that there would not be any Local Effect.

Determination of whether the software Failure Mode has a Safety Impact can only be made at the System (medical device) level. In the hierarchical multilevel FMEAs this can be known only after the integration of the FMEAs into the System DFMEA. But it may be possible to make some estimations of the Safety Impact in advance. For example, if it is certain that the Failure Mode would lead to one of the Hazards in the Clinical Hazards List, it would be a good guess that the Safety Impact will end up being Y. Another way to estimate the Safety Impact of a Failure Mode is if it would violate a System requirement which is tagged as Safety.

If the Safety Impact of the Failure Mode cannot be determined in advance, you can set the Safety Impact to N as a generic setting and use the "No-Safety Impact" column in the Ratings tab of the template to determine the Severity rating. As the SFMEA is a living process and goes through an iterative process, when the FMEAs are rolled up to the System DFMEA, it will become apparent whether a given Failure

Table 15.1 Definitions of Software Failure Modes and Effects Analysis severity ratings

Severity Criteria (Sev)

Rank	Qualitative criteria—no safety impact	Qualitative criteria—safety impact
5	*Catastrophic*: Described failure mode will cause immediate failure of the Subject. (Total loss of all functions—primary and secondary)	*Catastrophic*—Impact of the end-effect at the System level can be death
4	*Critical*: Described failure mode will severely impact Subject functionality | Complete loss of primary functions	*Critical*—Impact of the end-effect at the System level can be permanent impairment or life-threatening injury
3	*Serious*: Described failure mode will reduce Subject functionality. (Partial loss of primary functions | Complete loss of secondary functions)	*Serious*—Impact of the end-effect at the System level can be injury or impairment that requires professional medical intervention
2	*Minor*: Described failure mode will have temporal or self-restoring impact on functionality | Partial loss of secondary functions	*Minor*—Impact of the end-effect at the System level can be temporary injury or impairment that does not require professional medical intervention
1	*None*: Described component failure will have no impact on functionality	*Negligible*—Impact of the end-effect at the System level can be at most an inconvenience or temporary discomfort

Mode links up to any Hazards. After the integration of the FMEAs and creation of the System DFMEA, a cross-check is done to ensure consistency of Safety Impact ratings. Any End Effect that traces up to a Hazard must have a Y in the Safety Impact column.

Cite all the existing mitigations in the "Existing Mitigations" columns. Systemic Causes should be universally mitigated, and not repeated in every row. When estimating the ratings, assume the existing mitigations are implemented and effective.

There are three factors that are typically used to estimate the criticality of a Failure Mode: Severity, Occurrence, and Detectability.

Severity is the significance of the worst reasonable consequence of the End Effect at the boundary of analysis. Severity Ranking definitions are different depending on whether the End Effect has a safety impact or not. For End Effects that do not have a safety impact, use the left column in Table 15.1, and for those with a safety impact use the right column.

IEC 62304 [9] Annex B, section 4.4 states that unless a quantitative estimation of the probability of software failure is done, the probability for software failure should be presumed to be 1. This is true for systemic failures. However, in the SFMEA, we consider contribution of external factors to software failures. Therefore the Occ column would contain the likelihood that the software item could fail due to external factors. In the case of legacy software with available data for the probability of systemic software failure, Occ would compound the likelihood of software failure with the likelihood of the external factors. Let's say, e.g., that historical data shows that legacy software fails at a rate of 0.01% per 10,000 hours of operation. In the tank pressurization example earlier, if legacy data shows that the pressure sensor fails at a rate of 0.02% per 10,000 hours of operation, then the likelihood software failure due to a systemic failure *or* a sensor failure would be 0.03% per 10,000 hours of operation.

If Occurrence is based *only* on software failure rate, and there is no available data on software failure rate, then leave the Occ rating blank. Otherwise, use the qualitative, or quantitative guidelines in Table 15.2 to estimate a ranking for Occ.

Table 15.2 Definitions of Software Failure Modes and Effects Analysis occurrence ratings

Probability of Occurrence Criteria (Occ)

Category	Rank	Qualitative criteria	Quantitative criteria
Frequent	5	The occurrence is frequent. Failure may be almost certain or constant failure	$\geq 10^{-3}$
Probable	4	The occurrence is probable. Failure may be likely | Repeated failures are expected	$<10^{-3}$ and $\geq 10^{-4}$
Occasional	3	The occurrence is occasional. Failures may occur at infrequent intervals	$<10^{-4}$ and $\geq 10^{-5}$
Remote	2	The occurrence is remote. Failures are seldom expected to occur	$<10^{-5}$ and $\geq 10^{-6}$
Improbable	1	The occurrence is improbable. The failure is not expected to occur	$<10^{-6}$

Detectability in SFMEA has a similar connotation as in the DFMEA—it is an indication of how likely it is for the End Effect to be detected and countermeasures be taken, external to the boundary of analysis, to minimize the risk of Harm. This concept was elucidated in DFMEA workflow analysis, Section 12.6.1.3. A software Failure Mode with a safety impact is of lower criticality if it can be externally detected and countermeasures taken to minimize harm.

Internal detection and mitigations, such as CRC checks and error corrections are considered part of good design and serve to systemically reduce Occ ratings.

Refer to Table 15.3 for definitions of detectability rankings. Use quantitative data if available. Otherwise use the qualitative criteria to determine the Detectability rankings

Table 15.3 Software Failure Modes and Effects Analysis detectability ratings

Detection Criteria (Det)

Category	Rank	Qualitative criteria	Quantitative criteria
Undetectable	5	No detection opportunity | No means for detection | Countermeasures not possible	$<10^{-3}$
Low	4	Opportunity for detection is low | Countermeasures are unlikely	$<10^{-2}$ and $\geq 10^{-3}$
Moderate	3	Opportunity for detection is moderate | Countermeasures are probable	$<10^{-1}$ and $\geq 10^{-2}$
High	2	Opportunity for detection is high | Countermeasures are likely	$<9 \times 10^{-1}$ and $\geq 10^{-1}$
Almost Certain	1	Opportunity for detection is almost certain | Countermeasures are certain	$\geq 9 \times 10^{-1}$

Table 15.4 Software Failure Modes and Effects Analysis criticality table

RPN	Action
53–125	**Level 3**—Reduce RPN through failure compensating provisions.
13–52	**Level 2**—If Safety Impact is Y, reduce RPN to as low as possible. If Safety Impact is N, reduce RPN if feasible.
1–12	**Level 1**—If Safety Impact is Y, reduce RPN to as low as possible. If Safety Impact is N, further RPN reduction is not required.

Similar to DFMEA an RPN value is computed as the product of Sev, Occ, and Det ratings. Higher RPN indicates higher criticality. This number is used to prioritize the Failure Modes and determine the degree of compensation that must be exercised.

Table 15.4 offers a suggested stratification of compensating actions based on the criticality of the Failure Mode. The boundaries in Table 15.4 are selected at 12 and 52. But it is up to the manufacturer to decide where to draw the boundaries. Table 15.4 says that for the highest segment of RPN ratings, Level 3, the RPN must be reduced to a lower Level.

For Level 2, RPN should be reduced as far as possible, for safety-related Failure Modes. But for nonsafety-related Failure Modes, the decision as to how far to reduce the RPN is a business decision and depends on the feasibility of the actions needed to reduce the RPN.

Table 15.5 SDFMEA S-D Criticality Table

Criticality		Severity				
		1	2	3	4	5
Detectability	5	2	2	3	3	3
	4	1	2	2	3	3
	3	1	1	2	2	3
	2	1	1	1	2	3
	1	1	1	1	1	2

For Level 1, per EN ISO 14971:2012 [7] for safety-related Failure Modes the RPN must be reduced as far as possible, therefore the treatment of RPN is the same as for Level 2. However, for nonsafety-related Failure Modes, further action is not required.

If the Occ rating is left blank, then only Severity and Detectability rankings are available to determine the criticality of a software Failure Mode. In such cases, the template offers a two-dimensional criticality matrix (see Table 15.5). This criticality matrix also stratifies the software Failure Modes into three levels. The disposition of the three levels can follow the same action recommendations that are found in Table 15.4. The purpose of SFMEA is the identification of software-caused Hazards, and to prioritize software Failure Modes for mitigation. General mitigations such as static code checks, structured walkthroughs, and peer reviews benefit all of the software. But for software failures with high criticality, additional mitigations like external hardware, or external independent software mechanisms should be devised.

As with other FMEAs, SFMEAs serve two benefits: safety and reliability. For safety, all we need to determine is whether a software item can precipitate a Hazardous Situation. But for reliability, it is of interest to know the areas of software whose failure could impact product performance.

15.3 SOFTWARE SAFETY CLASSIFICATION

Manufacturers of medical devices that include software are required to assign a safety classification to the Software System, based on the potential risk of Harm to people from the Software System, in a worst-case scenario. According to Ref. [9],

"The Software System is software safety class A if:

- the software system cannot contribute to a Hazardous Situation; or
- the Software System can contribute to a Hazardous Situation which does not result in unacceptable risk after consideration of Risk Control measures external to the software system.

The Software System is software safety class B if:

– the software system can contribute to a Hazardous Situation which results in unacceptable risk after consideration of Risk Control measures external to the Software System and the resulting possible harm is non-serious injury.

The Software System is software safety class C if:

– the Software System can contribute to a Hazardous Situation which results in unacceptable risk after consideration of Risk Control measures external to the Software System and the resulting possible harm is death or serious injury."

Naturally these definitions necessitate the need to know what "serious Injury" is. IEC 62304 [9] sec. 3.23 defines Serious Injury as:

"injury or illness that:

a. is life threatening, or
b. results in permanent impairment of a body function or permanent damage to a body structure, or
c. necessitates medical or surgical intervention to prevent permanent impairment of a body function or permanent damage to a body structure

NOTE Permanent impairment means an irreversible impairment or damage to a body structure or function excluding trivial impairment or damage."

These definitions are not well aligned with the definitions of seriousness of Harm as offered in ISO 14971 [3,7] and cited in Table 17.1. One could roughly align the definition of Serious Injury with a combination of Critical and Serious from ISO 14971 [3].

All Harms have an associated Hazardous Situation. If the Hazardous Situation can manifest solely due to software failure, then P_1 for that Hazardous Situation is presumed to be 100% and risks are equal to the P_2 numbers for each Harm severity class. According to IEC 62304 [9] Section 7.2.2, the safety class of software that is intended to be a Risk Control is based on the risk that the risk-control-measure is controlling. If the software Risk Control is controlling the risk of a software item, then its class would be the same as the class of the *controlled software*. If the software Risk Control is controlling the risk of a hardware item, then use the following algorithm to determine the software safety class for the Risk Control software.

1. What Harm(s) can the hardware-failure cause? Utilizing the Harms Assessment List (HAL), identify the Harm severity class that has the highest probability. If more than one Harm is identified, choose the highest class among all the Harms.
2. Using Table 15.6 select the software safety class that correlates to the hardware severity class from Step 1.

Table 15.6 Harm severity versus software safety class

Harm severity class	Software safety class
Catastrophic	C
Critical	C
Serious	C
Minor	B
Negligible	A

Example: Consider Table 11.1. Imagine a hypothetical System in which a single hardware failure could cause both Harm.3 and Harm.4. In this System, there is a software Risk Control for that hardware failure. We want to know the safety class of the software Risk Control. The highest likelihood Harm for Harm.3 is the Minor class. The highest likelihood Harm for Harm.4 is the Serious class. Per Step 1 of the algorithm above, we select Serious class. Consulting Table 15.6, we see that Serious class correlates to Software Safety Class: C.

By this point, we have determined whether software has a contribution to safety, and have assigned a software safety class to it. If the software System performs many tasks, only some of which warrant a safety class of C or B, a manufacturer may desire to focus their engineering resources on the parts that are responsible for the safety critical functions. To do this requires the knowledge of the software architecture.

Only after the software architecture is completed, the full set of software items, and their contributions to safety can be known. Given the top-down System analysis that was done earlier, we can deduce which software items contribute to which Hazardous Situations.

If the software architecture and design strategy allows segregation and discrimination among the software Items, then a separate safety classification can be assigned to each software Item within the software System. If this strategy is chosen, segregation must be shown to be effective. That is, a mechanism that adversely affects one software item must not be able to adversely affect another, segregated software item. Examples of segregation strategies include use of separate microprocessors and partitioned memory. The adequacy/strength of the segregation strategy should be based on the level of risk involved.

The manufacturer may find that the cost and complexity of the segregation strategy is too high and instead opt to treat all of the software as the highest software safety class of the Software System.

Fig. 15.3, which is an adaptation of Figure 3 of IEC 62304 [9], provides a decision process for the determination of the safety risk class of the Software System. In the beginning, all software is assumed to be Class C by default. Following the process in Fig. 15.3, if software is initially determined to be class B or C, an external Risk Control measure is considered. This external Risk Control may be software or hardware. If the software option is chosen, doing a SFMEA

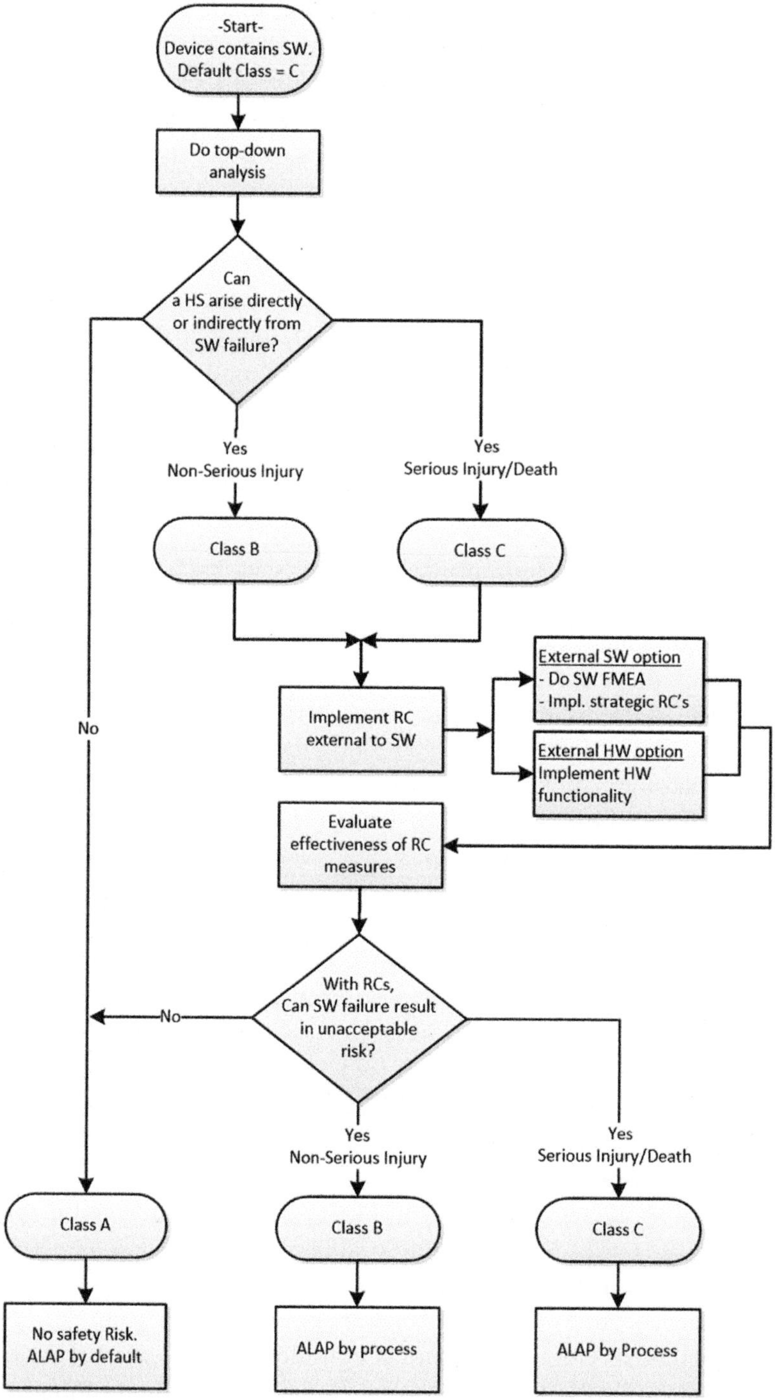

Figure 15.3 Software safety classification process.

allows the identification of the safety-critical software-items, and thus strategic application of Risk Controls to the highest risk software-items. It is also permissible to treat the software as a black box and apply the software Risk Controls to the entire software-system. If the hardware option is chosen, then appropriate functionality must be designed and implemented to be able to protect against the software failure.

After implementation of the external Risk Controls, the effectiveness of the Risk Controls is evaluated to judge the software contribution to overall device risk. If the software is still capable of producing unacceptable levels of risk, then the software development process should follow the different levels of rigor that are defined in IEC 62304 [9].

The software safety classification(s) and their rationale must be documented in the Risk Management File (RMF).

Risk Controls serve to prevent or reduce the probability of the occurrence of software-induced Hazardous Situations, or to reduce the severity of the ensuing Harms.

The Standard [9] clarifies that in addition to external hardware and independent software means, the external Risk Control could be a healthcare procedure, or even other means that can help reduce the contribution of software to the creation of Hazards.

15.4 THE BXM METHOD FOR SOFTWARE RISK ANALYSIS

In Section 15.1, two categories of software were identified: new software and legacy software. Per IEC 62304 [9], there is no consensus on a method to estimate the probability of occurrence of software failures in new software. Ref. [9] also acknowledges that for legacy software, it may be possible to quantitatively estimate the probability of occurrence of software failures, based on the usage of legacy software and examination of postproduction data. This creates two distinct pathways for software risk management:

Case 1. where the probability of software failure can be estimated—e.g., for legacy software.

Case 2. where the probability of software failure cannot be estimated—e.g., for new software, or legacy software for which postproduction data is unavailable, or is of questionable quality.

In this section, the BXM method for software risk analysis for each case is laid out.

15.4.1 Case 1—Legacy software

If postproduction data of adequate quality exists where the manufacturer can derive the probability of occurrence of software failures, then the estimation of risk due to software failures is possible and is similar to hardware risk analysis. That is, after the identification of software-caused Hazards, the probability of occurrence of the Hazard is computed by following the sequence of events as depicted in Fig. 3.1, where one of the events is a software failure with a known probability.

The risk estimation follows the quantitative method as described in Section 17.3.

15.4.2 Case 2—New software

For Case 2, the probability of occurrence of software failure cannot be estimated. As depicted in Fig. 3.1, in hazard theory, software failure would be an event in the chain of events leading to the realization of the Hazard, and the Hazardous Situation. Without knowing the probability of occurrence of the software failure, P_1 cannot be estimated. Following the guidance in ISO 14971 [3,7] section D.3.2.3, instead of trying to estimate P_1, we'll focus on reducing the probability of software failures, and thus reducing the related harms.

The steps of the BXM method for software risk analysis are:

1. *Ensure software requirements are correct.* Tools: modeling, simulation, reviews.
2. *Define the software architecture, and classify all software items* per IEC 62304 [9].
3. *Ensure software implementation is correct.* Tools: structured walkthroughs, peer reviews, testing, automation, use of robust processes, and levels of rigor prescribed for the different software safety classifications.
4. *Reduce safety classification of software items* (to the degree possible) via the use of Risk Controls that are external to the software.

Next, develop the software in compliance to IEC 62304 [9] and do not estimate the risks due to software-induced Hazards. It follows that without an estimate of software risks, software risks could not be included in the computation of the overall residual risks.

15.5 RISK MANAGEMENT FILE ADDITIONS

Implementation of IEC 62304 [9] introduces additional documentation requirements. The resulting artifacts are to be stored in the RMF. Table 15.7 lists these additional entries in the RMF.

15.6 RISK CONTROLS

Risk Controls fall in three categories:

1. Inherent safety by design
2. Protective measures
3. Information for safety

See further elaboration in Section 18.2 on Risk Control option-analysis.

Table 15.7 Additional documents for the risk management file

IEC 62304:2006 Reference	Additional software-related documentation entries to the risk management file	Safety class
4.3 c)	Software safety class assigned to the software system	N/A
4.3 d), e)	Software safety class for each software item, if the class of the software item is different from the class of the software item from which it was created by decomposition	N/A
4.3 f)	Rationale for using a lower software safety class for a software item in the software system	N/A
4.4.5	The version of the legacy software used, together with the rationale for the continued use of the legacy software	N/A
5.1.7	Software risk management plan	A, B, C
5.2.1	Software system requirements, as derived from the System-level requirements	A, B, C
5.2.6	Verification of the following: – software requirements implementation, and are traceability to parent System requirements – that software requirements are not contradictory – that software requirements are unambiguous – that software requirements are testable – that software requirements are unique	A, B, C
5.3.2	Architectural design of the interfaces among the software items and also interfaces between software items and external entities (hardware or software)	B, C
5.3.6	Verification of the software architecture to implement both System and software requirements, support the required interfaces, and support the proper operation of any SOUP items	B, C
5.4.2	Software design with sufficient detail to enable correct implementation of the software	C
5.4.3	Interface design among the software units and also interfaces between software units and external entities (hardware or software). Include sufficient detail to allow the correct implementation of the software	C
5.4.4	Verification of software detailed design to show that the software implements and doesn't contradict the software architecture	C
5.5.5	Verification of software units	B, C
5.6.3, 5.6.7	Software integration test results	B, C
5.7.5	Software system test cases, methods, tools, configurations, results, dates and identity of the testers	A, B, C
5.8.2	All known residual anomalies	A, B, C
5.8.4, 5.85	Versions of the medical device software that have been released, and the procedure and the environment required to create them	A, B, C

(*Continued*)

Table 15.7 (Continued)

IEC 62304:2006 Reference	Additional software-related documentation entries to the risk management file	Safety class
6.2.1.2, 9.2	Post-production monitoring of complaints, feedback, and documentation of the outcome of any investigation and evaluation.	A, B, C
7.1.4, 7.2.1	Potential causes of the software contribution to hazardous situations, and the relevant risk controls	B, C
7.3.3	Traceability from each hazardous situation to software item to software cause to risk control measure to verification of the risk control measure	B, C
8.1.2	If SOUP items are used, document the SOUP title, name of manufacturer and unique SOUP identifier	A, B, C
8.1.3	The configuration items and their version that comprise the software system configuration	A, B, C
9.5	Records of problem reports and their resolution including their verification	A, B, C

Risk Controls for direct software-caused Hazards would fall into one of the above three categories. Next are examples for each of the above three Risk Control:

1. *Inherent safety by design*: limiting software authority by hardware; e.g., physical limit to the maximum torque that a motor could deliver would mean that even if the software failed and gave a command for too much torque, it would be physically impossible for the motor to deliver that much torque.
2. *Protective measure*: algorithmic validity checking of inputs, password checks
3. *Information for safety*: warning messages, alarms

For indirect software-caused Hazards, certain unique Risk Controls apply. These are general strategies in software development such as pointer initialization, use of checksums on critical data, and avoidance of dynamic memory allocation. See Section 15.10 for many more tips for developing safety-critical software.

It is not practical to try to trace every potential software-caused Hazard to an indirect Cause and its mitigation. For example, a bit flip could have unpredictable effects anywhere in the software. Multiple software-caused Hazards could have the same indirect cause and the same Risk Control. The better strategy is to have a general Risk Control for the indirect Causes that applies throughout the software system, e.g., a periodic CRC check of the code space to detect any bit flips.

15.7 LEGACY SOFTWARE

Legacy software is defined as "medical device software which was legally placed on the market and is still marketed today, but for which there is insufficient objective evidence that it was developed in compliance with the current version of this standard" [9]. Note that legacy software is not the same thing as SOUP. See Section 15.8 to learn more about risk management of SOUP.

A manufacturer may intend to continue to use existing legacy software in their medical devices. In this case, objective evidence supporting the claim of safe continued-use of the legacy software is required. Evidence may be derived from comprehensive assessment of available postproduction field data. Sources of postproduction field data include:

- Complaints and feedback on the device
- Reported adverse events that are attributable to the device
- Anomalies that are found in-house during testing

When a new product uses legacy software, that software is still subject to IEC 62304 [9].

Legacy software may be used without change, or may be modified to create new software, or it may be integrated into a new software system. In case of modification or integration into a new software system, new risks may be introduced. Analysis must identify, estimate, and evaluate any potential additional risks.

To demonstrate compliance of legacy software with IEC 62304 [9] perform the following steps:

1. "Assess any feedback, including postproduction information, on legacy software regarding incidents and/or near incidents, both from inside manufacturer's own organization and/or from users."
2. Evaluate the legacy software for:
   a. integration in the overall device architecture
   b. continued validity of the Risk Control measures that are implemented in the legacy software
   c. any Hazardous Situations associated with the continued use of the legacy software
   d. any potential Causes that could induce Hazardous Situations via the legacy software
3. Define Risk Controls for each potential cause that could induce Hazardous Situations via the legacy software.
4. Perform gap analysis
   a. Examine the required deliverables per IEC 62304 [9] for the safety class of the legacy software.
   b. Compare what's required versus what's available. Where gaps are identified, evaluate the potential reduction in risk resulting from the generation of the missing deliverables and associated activities. Based on this evaluation, determine what additional deliverables and activities to perform. At a

minimum, Software System test records should be made available (see IEC 62304 [9] sec 5.7.5).

Note—Ensure that any Risk Controls that are implemented in the legacy software are included in software requirements.

c. Assess the continuing validity of the available deliverables.

d. Evaluate the adequacy of existing risk management documentation vis-à-vis ISO 14971.

15.8 SOFTWARE OF UNKNOWN PROVENANCE

Software of Unknown Provenance or Unknown Pedigree (SOUP) refers to software that is obtained from a third party, for which adequate documentation and records of development process is not available. SOUP is of particular interest in areas where software plays a pivotal role in the safety of the System.

Most medical device software uses some SOUP, as it is not practical to produce software from scratch. So, managing the risk of SOUP becomes a fact of life.

FDA Guidance [27] states that "It may be difficult for you to obtain, generate, or reconstruct appropriate design documentation as described in this guidance for SOUP. Therefore, we recommend that you explain the origin of the software and the circumstances surrounding the software documentation. Additionally, your Hazard Analysis should encompass the risks associated with the SOUP regarding missing or incomplete documentation or lack of documentation of prior testing. Nonetheless, the responsibility for adequate testing of the device and for providing appropriate documentation of software test plans and results remains with you."

IEC 62304 [9] requires that the software configuration, integration, and change management plan include SOUP.

With respect to risk management, the functional and performance requirements of SOUP, and hardware and software that is necessary for the proper function of SOUP must be identified.

If failure or unexpected results from SOUP could potentially contribute to a Hazardous Situation, at a minimum evaluate any anomaly list that is published by the supplier of the SOUP to determine if any of the known anomalies could potentially create a System Hazard, or lead to a Hazardous Situation.

15.9 SOFTWARE MAINTENANCE AND RISK MANAGEMENT

For most medical devices, software is continuously updated over time. The reasons could be bug fixes, feature improvement, or cybersecurity fixes. Changes to software can disrupt existing Risk Control measures, and/or introduce new Causes and

Hazards. Formal and effective Quality management is essential to control software changes and assess their impacts.

One of the issues for software maintenance releases is that in many cases the team that produces the maintenance-releases is not the same as the team that produced the original software. As such they may be unfamiliar with the rationales for the original work, or Risk Controls that are in place.

Strategies to reduce risks due to software maintenance include: good documentation and organizational strategies, e.g., to keep some of the staff who worked on the original software, available for consultation.

15.10 SOFTWARE RELIABILITY VERSUS SOFTWARE SAFETY

A fact to consider is that reliable software is not necessarily safe. Reliability is defined as the ability to deliver the intended function for a certain length of time, under certain operating conditions. Software that is designed to the wrong requirements could be implemented with no defects, operate as intended 100% of the time, and still be unsafe. Conversely, unreliable software may be safe. Consider an automatic sphygmomanometer (blood pressure monitor) (Fig. 15.4). Imagine that this device has an overpressure sensor that would detect if the device applies too high a pressure to the cuff, possibly injuring the patient. Now consider a condition where the control-software incorrectly interprets the overpressure sensor and unnecessarily deactivates the device as a safety Risk Control. In this example, unreliable software creates annoyance to the user, but does not create a safety Hazard.

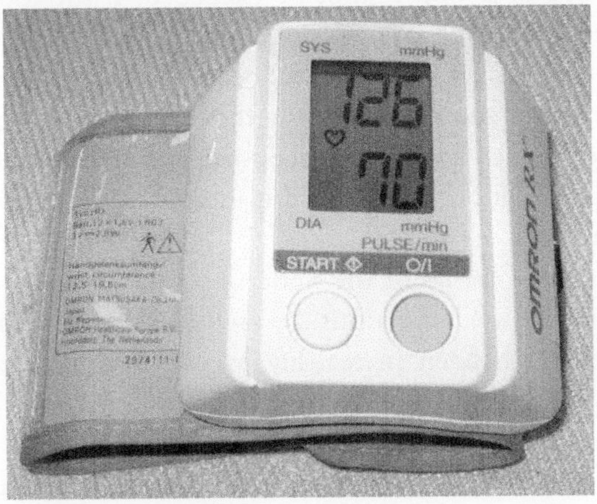

Figure 15.4 Automatic sphygmomanometer.

15.11 Tips for developing safety-critical software

In this section, general tips and advice are provided to aid in the successful development of safety-critical medical software.

- Do not postpone software risk analysis until late in the product development process. Retrospective software risk analysis cannot effectively reduce risk.
- When adding new software features, analyze for introduction of new Hazards, or compromising existing software Risk Controls.
- Consider platform evolution, e.g., operating system upgrades in your risk management process.
- Ensure that good software configuration management and safety impact analysis due to software changes are a part of your Quality Management System. This would help detect changes with unexpected consequences.
- If possible, separate the safety-critical software items and keep them as simple and small as possible.
- IEC TR 80002-1 [26] table C.1 offers advice on avoiding pitfalls in the development of software for medical devices.

NASA JPL lead scientist, Gerard J. Holzmann has produced a set of 10 rules to help guide the development of safety-critical software. Below, the 10 rules are summarized. For further details, see Ref. [28]. These rules are directed at C programming language, but you can benefit from them in other programming environments as well.

1. *Rule*: Restrict all code to very simple control flow constructs — do not use *goto* statements, *setjmp* or *longjmp* constructs, and direct or indirect *recursion*.
2. *Rule*: All loops must have a fixed upper-bound. It must be trivially possible for a checking tool to *prove* statically that a preset upper-bound on the number of iterations of a loop cannot be exceeded. If the loop-bound cannot be proven statically, the rule is considered violated.
3. *Rule*: Do not use dynamic memory allocation after initialization.
4. *Rule*: No function should be longer than what can be printed on a single sheet of paper in a standard reference format with one line per statement and one line per declaration. Typically, this means no more than about 60 lines of code per function.
5. *Rule*: The *assertion density* of the code should average to a minimum of two assertions per function. Assertions are used to check for anomalous conditions that should never happen in real-life executions. Assertions must always be side-effect free and should be defined as Boolean tests. When an assertion fails, an explicit recovery action must be taken, e.g., by returning an error condition to the caller of the function that executes the failing assertion. Any

assertion for which a static checking tool can prove that it can never fail or never hold violates this rule. (I.e., it is not possible to satisfy the rule by adding unhelpful "*assert(true)*" statements.)

6. *Rule*: Data objects must be declared at the smallest possible level of scope.
7. *Rule*: The return value of non-void functions must be checked by each calling function, and the validity of parameters must be checked inside each function.
8. *Rule*: The use of the preprocessor must be limited to the inclusion of header files and simple macro definitions. Token pasting, variable argument lists (ellipses), and recursive macro calls are not allowed. All macros must expand into complete syntactic units. The use of conditional compilation directives is often also dubious, but cannot always be avoided. This means that there should rarely be justification for more than one or two conditional compilation directives even in large software development efforts, beyond the standard boilerplate that avoids multiple inclusion of the same header file. Each such use should be flagged by a tool-based checker and justified in the code.
9. *Rule*: The use of pointers should be restricted. Specifically, no more than one level of dereferencing is allowed. Pointer dereference operations may not be hidden in macro definitions or inside *typedef* declarations. Function pointers are not permitted.
10. *Rule*: All code must be compiled, from the first day of development, with *all* compiler warnings enabled at the compiler's most pedantic setting. All code must compile with these setting without any warnings. All code must be checked daily with at least one, but preferably more than one, state-of-the-art static source code analyzer and should pass the analyses with zero warnings.

CHAPTER 16

Integration of Risk Analysis

Abstract

Once the System is decomposed, it is possible to perform a hierarchical multilevel Failure Modes and Effects Analysis (FMEA) on the System components. Lower level components' FMEAs, as well as input from supplier FMEAs roll up into higher level FMEAs. This process continues until the System Design Failure Modes and Effects Analysis from which System hazards can be derived.

Keywords: Integration; multi-level FMEA; supplier risk management

16.1 HIERARCHICAL MULTILEVEL FAILURE MODES AND EFFECTS ANALYSIS

Once the System is decomposed, it is possible to perform a hierarchical multilevel Failure Modes and effects analysis (FMEA) on the System components. Lower level components' FMEAs roll up into higher level FMEAs. This process continues until the L1, System Design Failure Modes and Effects Analysis (DFMEA) (Reference—Fig. 12.5).

Fig. 16.1 illustrates the connectivity between consecutive decomposition levels. The rationale for this construct is that lower level components can fail due to design issues or manufacturing issues. From the perspective of an upper level DFMEA, what matters is that the lower level component has failed. Whether it was due to a design or a manufacturing issue is only relevant when the Cause of that failure is documented.

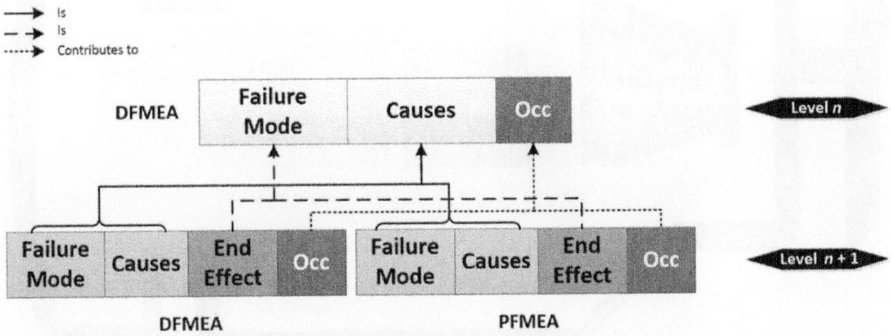

Figure 16.1 Failure Modes and Effects Analysis integration.

Safety Risk Management for Medical Devices
DOI: https://doi.org/10.1016/B978-0-12-813098-8.00016-7

© 2018 Elsevier Ltd.
All rights reserved.

This concept is elucidated below in an example of an automobile System. An automobile is comprised of several subsystems (see Fig. 16.2). For this example, let's consider the propulsion system—the Engine.

In the illustration in Fig. 16.2 the Engine is a black box. But the Engine itself is comprised of many other subsystems. In Fig. 16.3 a number of Engine subsystems are illustrated. Let's consider the cooling system—the Water Pump.

In the illustration in Fig. 16.3 the Water Pump is a black box. But the Water Pump itself is comprised of many other parts. In Fig. 16.4 the Water Pump is decomposed to its constituent parts. We will not further decompose the Water Pump parts.

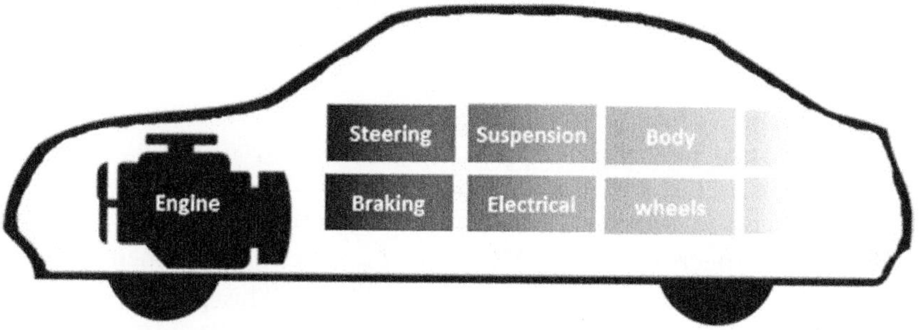

Figure 16.2 Automobile subsystems.

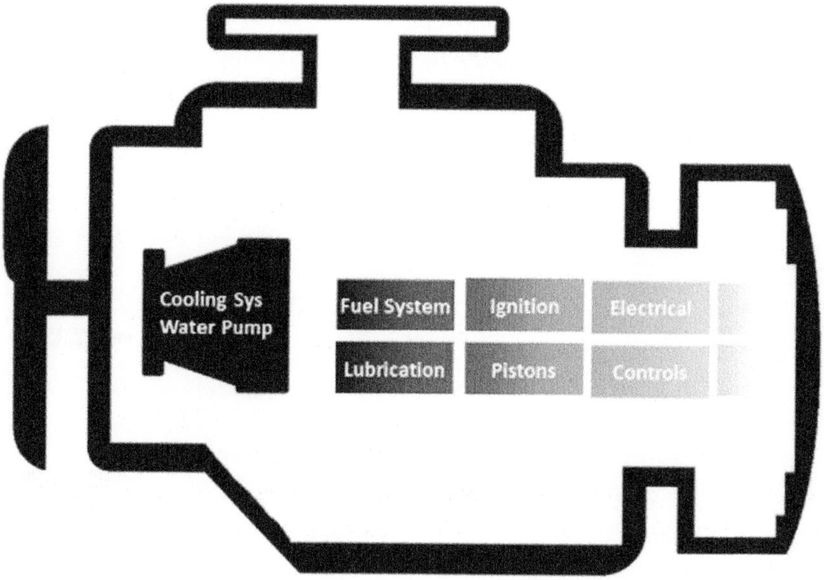

Figure 16.3 Engine subsystems.

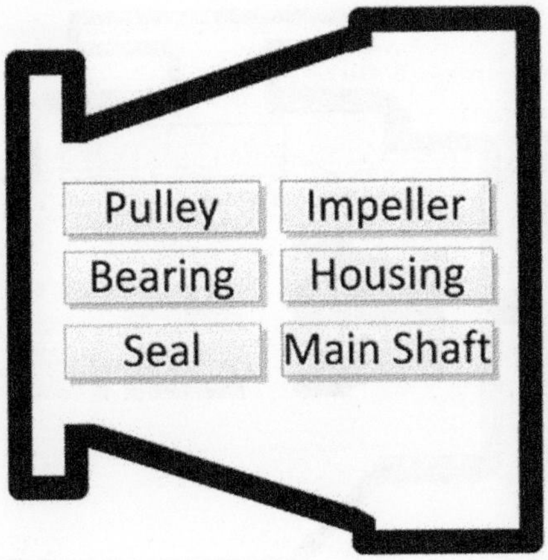

Figure 16.4 Water Pump.

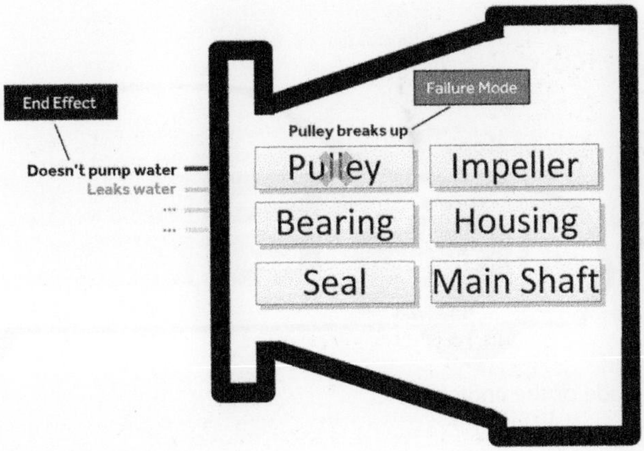

Figure 16.5 Failure Mode of the Pulley.

To this point, we have followed a decomposition model such as what is illustrated in Fig. 10.1. Now let's consider the Failure Mode of one of the Water Pump parts: the Pulley (Fig. 16.5).

If the Pulley breaks up, the water pump can no longer receive torque from the engine, and it stops pumping water. Not pumping water is the End Effect of the pulley break-up on the Water Pump. Now let's move up one level to the Engine.

To the Engine, the Water Pump is a black box component, whose Failure Mode is: *Doesn't Pump Water.* Notice that what was the End Effect: *Doesn't Pump Water* from the Water Pump perspective, has become the *Failure Mode* of the Water Pump from the engine perspective, which sees the water pump as a black box component (Fig. 16.6).

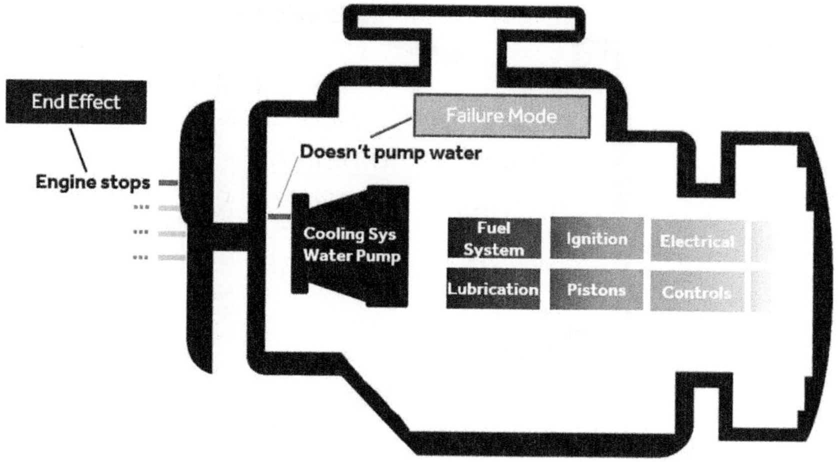

Figure 16.6 Failure Mode of the Water Pump.

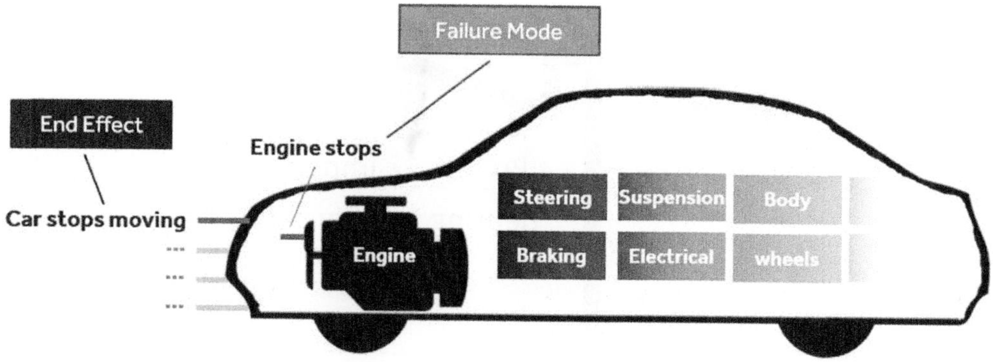

Figure 16.7 Failure Mode of the engine.

The same construct can continue upward to the car-level, where the End Effect of *Engine Stops* at the engine level can become the Failure Mode of the engine from the perspective of the car. Again, what was the End Effect: *Engine Stops* from the engine perspective, becomes the Failure Mode of the engine from the car perspective, which sees the engine as a black box component. Fig. 16.7 illustrates the engine as a black box from the perspective of the car.

16.2 INTEGRATION OF SUPPLIER INPUT INTO RISK MANAGEMENT

Most medical device manufacturers use suppliers to produce parts of their medical devices. The design and manufacturing of the supplied parts plays a role in the safety of the medical device. The input from the suppliers must be incorporated in the risk management of the medical device.

At a minimum, what is needed from suppliers are the Failure Modes of the products that they supply, and the occurrence ratings of the Failure Modes. With that information, we can incorporate the performance of the supplied product in the safety performance of the System. Occurrence rating helps us derive P1. Knowledge of Causes of the Failure Modes can help with estimation of P1s, and also with design of Risk Controls. Recognize that what is the Failure Mode to us, is the End Effect to the supplier's FMEA.

Knowledge of other entries in the supplier risk management file would be interesting for R&D engineering, Supply chain, and Quality departments, but not required from the risk management perspective.

CHAPTER 17

Risk Estimation

Abstract

According to ISO 14971 risk is a combination of the probability of occurrence of Harm and the severity of that Harm. Three methods are presented for risk estimation: qualitative, semi-quantitative, and quantitative.

Keywords: Risk estimation; qualitative; semi-quantitative; quantitative

According to ISO 14971 [3,7] risk is a combination of the probability of occurrence of Harm and the severity of that Harm.

Three methods are commonly used to estimate the risk of Harm. In increasing order of preference, they are: Qualitative, Semi-quantitative, and Quantitative. Each method is described below.

17.1 QUALITATIVE METHOD

The qualitative method is used when quantifiable data is unavailable, or confidence in the available data is low. In such cases use an N × M matrix such as the example in Fig. 17.1 to stratify the risks. In this example a 3 × 3 matrix is used to stratify the risks into three zones: high (*red*), medium (*yellow*) and low (*green*).

In order for this method to work, very good definitions for each probability and severity level should be given to ensure repeatability and consistency of ratings by different analysts, at different times. Tables 17.1 and 17.2 offer examples of language that could be used to promote consistency in severity and probability ratings.

17.2 SEMIQUANTITATIVE METHOD

The semiquantitative method is similar to the qualitative method, but with the difference that data is available for probability of occurrence of Harm. Generally, this is true for products that have been in the field for a significant length of time and about which a lot of field data has been collected.

The scales for probability of occurrence of Harm would be different for different products. Examples: "probability of harm per use," "probability of harm per device," "probability of harm per hour of use."

Safety Risk Management for Medical Devices
DOI: https://doi.org/10.1016/B978-0-12-813098-8.00017-9

© 2018 Elsevier Ltd.
All rights reserved.

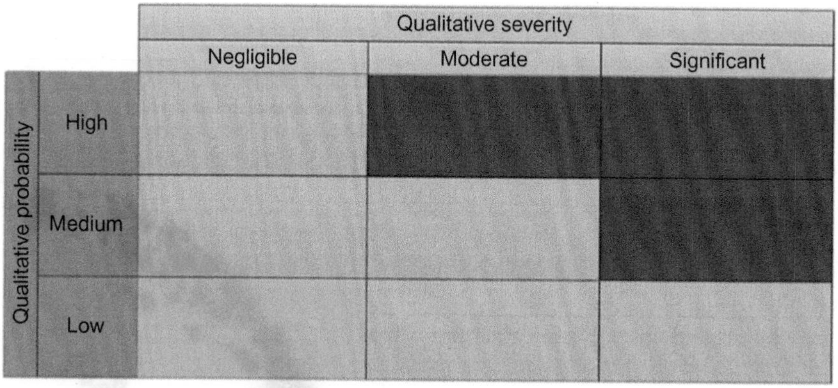

Figure 17.1 Example 3 × 3 qualitative risk matrix.

Table 17.1 Example three-level definitions for severity

Term	**Definition**
Significant	Death or permanent impairment/injury
Moderate	Reversible or minor injury
Negligible	Discomfort or inconvenience

Table 17.2 Example three-level definitions for probability

Term	**Definition**
High	Likely to happen | often | frequent
Medium	Can happen, but not frequently
Low	Unlikely to happen | rare | remote

Each Hazard would have an estimated risk based on the available data for the probability of occurrence of Harm, and the estimated severity of that Harm.

ISO 14971 [3,7] offers a 5 × 5 example for ranking severity and probability. These examples are cited in Tables 17.3 and 17.4.

Table 17.3 Example five-level definitions for severity

Rank	**Term**	**Definition**
5	Catastrophic	Death
4	Critical	Permanent impairment or life-threatening injury
3	Serious	Injury or impairment requiring professional medical intervention
2	Minor	Temporary injury or impairment not requiring professional medical intervention
1	Negligible	Discomfort or inconvenience or no harm

Table 17.4 Example five-level definitions for probability

Rank	Term	Definition
5	Frequent	$\geq 10^{-3}$
4	Probable	$< 10^{-3}$ and $\geq 10^{-4}$
3	Occasional	$< 10^{-4}$ and $\geq 10^{-5}$
2	Remote	$< 10^{-5}$ and $\geq 10^{-6}$
1	Improbable	$< 10^{-6}$

Fig. 17.2 depicts an example 5 × 5 risk matrix which stratifies the risks in the: high (*red*), medium (*yellow*), and low (*green*) zones.

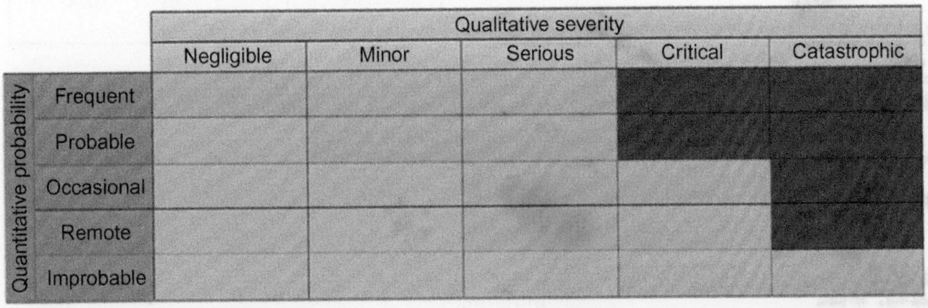

Figure 17.2 Example 5 × 5 risk matrix.

17.3 QUANTITATIVE METHOD

ISO 14971 [3] Annex E presents a concept for quantification of risk. This concept is presented in Fig. 17.3, where it is indicated that risk is the product of P_1, the probability of occurrence of the Hazardous Situation, and P_2, the probability of occurrence of Harm. The problem with this method is that manufacturers are forced to conservatively consider the worst Harm that could result from the Hazardous Situation. Clearly, the worst Harm doesn't happen every time. Considering the worst Harm distorts risk management in two ways:

1. If the worst Harm is unlikely to happen, but a moderate Harm is likely to happen, the worst-case risk of Harm would show a smaller risk, where as if the moderate Harm probability was considered, a higher risk number would have been computed. The result of this distortion is that the manufacturer would improperly rank the true risks of the medical device
2. Most Harms could result in death—at least with a small probability. Estimating risks based on the worst outcome for every Harm would create an illusion that the medical device is deadly. This could create an impossible situation for the manufacturer because unless the Hazard is eliminated, no matter how much P_1 is reduced, there would still be a risk of death, since P_2 is nonzero.

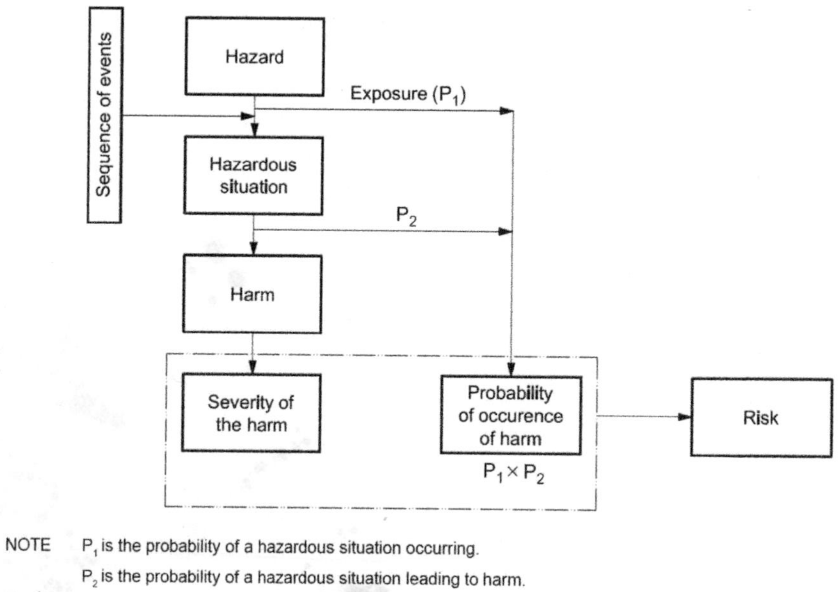

Figure 17.3 ISO 14971, Fig E.1—Quantification of risk.

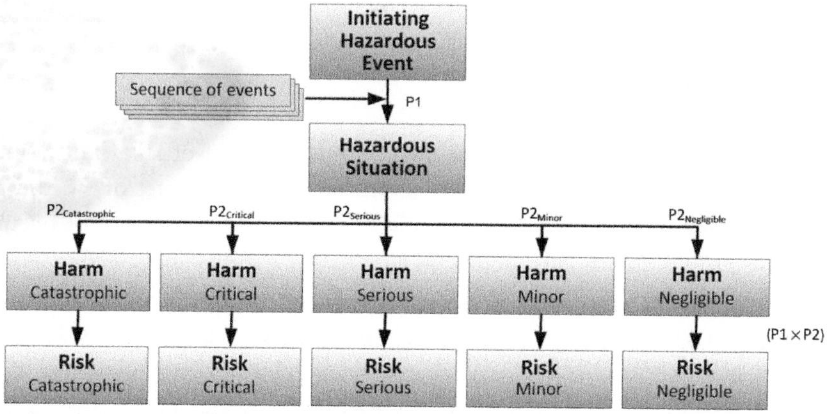

Figure 17.4 The BXM five-level risk computation method.

The BXM method uses a quantitative method that computes risk in five classes of Harm severity: catastrophic, critical, serious, minor, and negligible. This is depicted in Fig. 17.4. The advantage of this method is that the entire spectrum of Harm severities is considered and regardless of the severity of the Harm, the highest risk is identified.

Vis-à-vis the hazard theory that was presented in Section 3.2, since the accounting for Harm encompasses the entire spectrum from death to no-harm, it can be said that

P_1 = Probability of occurrence of Hazardous Situation = Probability of occurrence of Harm.

As in the semiquantitative method, P_1 is presumed to be known.

Consider this hypothetical example: Based on field data it is known that patients are exposed to too-high X-ray radiation in 1 out of 100,000 (10^{-5}) applications of an X-ray machine.

Now, given that a patient *is* exposed to too high X-ray radiation, we can ask:

- What is the probability that the patient experiences catastrophic Harm; $P2_{CATASTROPHIC}$
- What is the probability that the patient experiences critical Harm; $P2_{CRITICAL}$
- What is the probability that the patient experiences serious Harm; $P2_{SERIOUS}$
- What is the probability that the patient experiences minor Harm; $P2_{MINOR}$
- What is the probability that the patient experiences negligible Harm; $P2_{NEGLIGIBLE}$

In other words, given all the reported cases of X-ray overexposure, how many died; how many suffered permanent life-threatening injury; and so on. This exercise produces five P_2 numbers.

In this hypothetical example, we have P_1, the probability of X-ray over exposure (10^{-5}), therefore we can compute the risk of Harm from X-ray over exposure in five severity classes:

Risk of catastrophic Harm (death) = $P_1 \times P2_{CATASTROPHIC}$
Risk of critical Harm = $P_1 \times P2_{CRITICAL}$
Risk of serious Harm = $P_1 \times P2_{SERIOUS}$
Risk of minor Harm = $P_1 \times P2_{MINOR}$
Risk of negligible Harm = $P_1 \times P2_{NEGLIGIBLE}$

By repeating this process for all the Hazards that a given medical device presents, we can estimate the overall risk of Harm from the medical device.

It should be noted that all statistics are approximations of the truth. Therefore since both P_1 and P_2 are estimates, then Risk (R) is also an estimate.

In practice, we find that field-data may not be available for all P_1s and P_2s. In such cases, we will have to rely on alternative means of data gathering, such as published scientific papers.

Tip –"Severity" as used in *risk estimation* refers to the impact on health and safety. Severity as used in the *FMEA* refers to the significance of the worst reasonable consequence of the End Effect at the boundary of analysis, which might be unrelated to safety. Don't confuse the two.

17.4 PRE-/POST-RISK

Although ultimately what matters is the level of risk *after* the implementation of Risk Controls, traditionally, some members of the risk management community look for risk estimations *before* and *after* the implementation of Risk Controls.

Experienced medical device designers apply their knowledge of medical devices to their preliminary designs to ensure that the product is as safe as possible. Therefore it is entirely possible that the initial design of the product is the best and final design. ISO/IEC Guide 51 [1] says "During the preliminary design of a product or a system, inherently safe design measures are usually intuitively applied. Therefore, the risk evaluation for some hazards might lead to a positive outcome at the first iteration and no further risk reduction is required."

Determination of risk, pre-Risk Controls is sometimes not practical. For example, let's say the designers specified good, biocompatible materials for an implantable device from the preliminary stages of design. To answer the question of what was the risk before choosing biocompatible materials, would require one to hypothesize the use of bio-incompatible materials. But which material? There are a range of toxic materials that could be selected. And what benefit would this exercise deliver?

Another dilemma is the definition of *post*-Risk Control. Consider a first draft of a design that creates an unacceptable risk. The design is improved with some Risk Controls, and a second draft is created. The risk of the second draft also is judged to be unacceptable. A third draft is created with yet more Risk Controls. Now the risk of the third and final draft of the design meets the acceptability criteria. Which is the *post*-risk? Risk of draft 2, or risk of draft 3?

There are two ways to handle this matter.

1. Suffice it to provide the risk of the final design, as *that* is what matters to the patient.
2. Take the risk of the first draft of the design and call it pre-risk. Keep it static. If this risk is acceptable, copy it in the post-risk column, so the pre- and post-risks are the same. If the pre-risk is not acceptable, then populate the post-risk column with the risk of the *final* design, regardless of how many iterations occur between draft 1 and final draft.

CHAPTER 18

Risk Controls

Abstract

Risk Controls are the overt actions and measures by which risks are reduced to, or maintained within, specified levels. Three types of risk control measures are presented, and risk control option analysis is discussed. Also, the concept of single-fault-safety is also expounded in this chapter.

Keywords: Risk controls; risk control option analysis; information for safety; completeness of risk controls; single-fault-safe

Once the risks of a medical device are estimated, measures must be taken to reduce the risks As Far As Possible [7]; or, if you are conforming to ISO 14971:2007, the risk are to be reduced to As Low As Reasonably Practicable. These measures are called Risk Controls.

Risk Controls can be viewed over two horizons:

1. Risk Controls performed prior to release of the product
   These Risk Controls are discussed in Section 18.2.
2. Risk Controls performed after the release of the product
   These are Risk Controls that are done at the customer site. Examples: personal protective equipment, organizational procedures, and training.

In general, Risk Controls attempt to prevent the realization of Hazards, or exposure to Hazards. These types of Risk Controls reduce P1. Some Risk Controls attempt to reduce the severity of the Harm after exposure to Hazards. These types of Risk Controls reduce P2. For example, antilock brakes reduce the probability of collision and impact by a car, but airbags reduce the severity of injury if a collision occurs.

18.1 SINGLE-FAULT-SAFE DESIGN

ISO 14971 [3,7] requires that the device risks under both normal and fault condition be managed. IEC 60601-1 [8] requires that medical devices be designed such that they are single-fault-safe. IEC 60601-1 [8] §4.2.2 further clarifies that "fault condition" includes single-fault condition, but is not limited to it. The concept of single-fault-safe has a built-in assumption of independence of faults. If the occurrence of the initial fault will necessarily cause the occurrence of a secondary fault, then they count as one fault.

Safety Risk Management for Medical Devices
DOI: https://doi.org/10.1016/B978-0-12-813098-8.00018-0

© 2018 Elsevier Ltd.
All rights reserved.

For example, if the failure of a device's user interface (fault #1) will certainly lead to the inability of the user to operate the device (fault #2), then these count as one fault.

A common interpretation of "single-fault-safe" is that as long as a medical device is safe under a single fault condition, the device risks are acceptable. But in fact, this is not true. Consider a device that can fail due to a single fault, and whose failure creates an unsafe condition. Assume the likelihood of occurrence of the single fault is high. Now a secondary means of protection is added, such that when the primary fault happens, the secondary means would transition the device to a safe state. Theoretically this device is single-fault-safe because it takes two independent faults to create an unsafe condition. But what if the likelihood of failure of the secondary means is also high? Can you envisage a situation where both the primary fault and the failure of the secondary means have occurred simultaneously? Given the knowledge that the likelihood of both the primary fault and the failure of the secondary means is high, you can surmise that the safety risk of the device would not be low.

A closer scrutiny of IEC 60601-1 [8] §4.7 reveals that Ref. [8] accepts a single means of risk reduction as single-fault-safe, if the probability failure of that single means is negligible. In the designs where arriving at an unsafe condition requires two faults, Ref. [8] clarifies that single-fault-safe is met, if the initial fault is detected before the secondary fault has occurred. Single-fault-safe is also met if the probability of failure of the secondary means is negligible, during the expected service life of the device. From the risk management perspective, what matters is that the overall residual risk of the device be acceptable, irrespective of one, two, or more faults.

In summary, the risk of the medical device must be acceptable during the mission of the device. Mission could be the expected service-life of the device. Or, if routine maintenance is done during which the failure of the secondary means would be detected, mission would be the time between maintenance events. Note the assumption of detection of failure of the secondary means, and the implicit repair/replacement of a failed secondary means during the maintenance event.

With this interpretation, we can compute the device risk based on the probability of occurrence of both the primary fault and the failure of the secondary means of protection during the mission of the device.

18.2 RISK CONTROL OPTION ANALYSIS

The Standard [3,7] identifies three methods of controlling risk as listed below in decreasing order of preference.

1. Inherent safety by design
2. Protective measures in the medical device itself or in the manufacturing process
3. Information for safety

The manufacturer is required to consider all the options and implement as many of the options as possible [7].

Consider if it is possible to eliminate a Hazard. If so, change the design so that the device is inherently safe from that Hazard. If elimination of the Hazard is not possible, then consider if protective means in the design, or the manufacturing process could protect the patient/user from Harm. Additionally, if providing information for the safe operation and use of the device could help with reducing the risks of the device, provide such information.

Document the Risk Control option-analysis and the decisions made on the selection and implementation of the Risk Controls.

After the first pass through risk analysis, it may be determined that additional Risk Controls need to be implemented. For every additional Risk Control that is implemented determine if any new Hazards are introduced or if the any of the current risks are increased.

18.3 DISTINCTIONS OF RISK CONTROL OPTIONS

Sometimes it may not be easy to identify the type of a particular Risk Control measure: inherently safe by design, or a protective measure, or information for safety. The different options are distinguished as follows:.

Safe by Design—The device behaves in a safe manner without any action or knowledge required by the user. The user cannot easily defeat the Risk Controls.

Example: Elevators have an automatic braking system that arrest the movement of the elevator in the case of mechanical or electrical failures. The user of the elevator doesn't need to know anything about the workings of the elevator, *and* cannot easily defeat the Risk Control.

Example: In a car with automatic transmission, it is possible to start the car only if the gear shift selector is in Park, or Neutral positions. This prevents the operator from starting the engine in gear and causing possible unexpected vehicle movement.

Protective Measure—Device behaves in a safe manner without the need for user intervention, but the protective measure(s) can be easily defeated by the user.

Example: Hypodermic needles come with a protective cap. The user does not need to take any action to make the product safe, as it is delivered. However, the user can easily remove the protective cap.

Information for Safety—Provides knowledge to the user and requires action by the user.

Example: Instruction for cleaning and sterilization of a reusable surgical tool provides knowledge to the user and requires action from the user in order to use the device safely.

18.4 INFORMATION FOR SAFETY AS A RISK CONTROL MEASURE

Information for safety can take many forms. For example: screen displays, IFUs, labels attached to the device, and online help.

The release of EN ISO 14971:2012 [7] created a lot of confusion and controversy, particularly with respect to Annex ZA, section 7, which states: "... information given to the users does not reduce the (residual) risk any further."

The problem was that in some cases there is no other option but to inform the user on how to use a device safely. Example: Reusable surgical instruments need to be cleaned and sterilized before each use. Since this action is performed by the user, providing information on how to clean and sterilize the device is the main way to ensure the safe reuse of the surgical instruments.

In 2014 the Notified Bodies Recommendation Group (NBRG) released a Consensus Paper [22] that provided guidance to Notified Bodies on how to interpret and apply Annexes Z in EN IS 14971:2012 [7]. In this Consensus Paper a distinction is made between *Disclosure of Residual Risk* and *Information for Safety.*

Disclosure of residual risks	Information in the accompanying documents on risks remaining after all Risk Control measures have been taken
Information for safety	Instructions of what actions to take or to avoid in order to prevent a Hazardous Situation from occurring

The Consensus Paper [22] advises the Notified Bodies that: "Any information for safety comprising instructions of what actions the user can take or avoid, in order to prevent a Hazardous Situation from occurring may be considered a risk control measure."

Disclosure of the residual risk enables a user to make an informed decision as to whether to use a medical device. Whereas, information for safety enables a user to safely use a device *after* he/she has decided to use the medical device.

The disclosure of the residual risk can also be used by the user of the medical device to better prepare for possible side effects or hazards that can occur during or after the use of the medical device.

18.4.1 Criteria for information for safety

If you choose to use information for safety as a Risk Control, there are certain considerations. The information for safety must be perceivable, comprehensible, and actionable by the user, *and* be effective in reducing risk.

Guide 51 [1] states "The content of an instruction should provide product users with the means to avoid harm caused by a product hazard that has not been eliminated or reduced, enable product users to make appropriate decisions concerning the use of the product..."

When using information for safety as a Risk Control, consider the following:

To whom—To whom will you be communicating? A trained clinician? A home user? Elderly? Youth? How well is it perceived and comprehended?

How—Will you be using words? Icons? What type of media (printed, digital screen)? What location/timing of the information? What level of detail?

What—What are the hazardous conditions? What are the consequences of exposure and what should be done to prevent Harm? In what priority should actions be taken?

Information for safety can be in many forms. For example: in user manuals, in labeling that is attached to the medical device, in graphical user interfaces such as screens, or even in the form of audio or tactile annunciations.

The primary intention should be avoidance of Hazardous Situations, either by prevention of Hazards, or by prevention of exposure to Hazards. Secondarily, information for safety should offer guidance on remedial actions if Harm has happened. The priority of actions should be commensurate with the level of risk, and be properly communicated to the users. Use of word such as: "Danger," "Warning," "Caution," "Alert," and "Note" could indicate the priority of the information for safety.

The three guiding principles in the utilization of information for safety as Risk Control are:

Inform | Motivate | Enable

According to Guide 51 [1] labeling should be:

- "conspicuous, legible, durable and understandable;
- worded in the official language(s) of the country/countries where the product or System is intended to be used, unless one of the languages associated with a particular technical field is more appropriate;
- concise and unambiguous."

With respect to legibility, consider the user population. What visual acuity can be expected? Under what lighting conditions will the user typically use the device? Use AAMI HE75 [20] for guidance on font sizes for visual displays.

The labeling should be durable and not fade, rub off, smear, or separate from its intended location during the expected life of the device.

Whether in audio or visual format, the information for safety should be understandable by the users. In today's global economy, with so many languages and cultures this is a challenging requirement. Even the choice of colors may convey one thing in one culture, and another in a different culture. For example, in the American culture the color yellow is usually an indication of a warning. But in the Japanese culture yellow is an indication of sickness and ill health.

Translations are a sensitive and critical aspect of communication of information for safety. In many companies the information for safety is drafted in a central language, usually English, and then translated to other languages. The problem is that not all concepts can be directly translated from one language to another. Here are some examples of potential sources of confusion in translations:

English: Bend the wire into a J, or hockey stick shape
Problem: Many languages don't have the letter J. Many cultures are unfamiliar with hockey.

English: When the alarm goes off do . . .
Problem: Does this mean when the alarm begins, or when the alarm ends?

Sometimes a translated word has one connotation in one geography, and another in a different geography. French in France is not the same as French in Canada; Portuguese in Portugal is not the same as Portuguese in Brazil, and English in the United States is not the same as English in the United Kingdom. For example:

First Floor
United States: Ground floor
United Kingdom: the floor above the ground floor

Chips
United States: Thinly sliced, deep-fried, baked, or kettle-cooked crunchy potatoes (crisps in the United Kingdom)
United Kingdom: Cut and deep-fried potatoes

Another element of the guidance from Guide 51 [1] is conciseness and unambiguity. This is a challenge for technical writers, particularly with respect to product labeling and screen user interfaces where space is limited.

As stated earlier, an important requirement for information for safety is that it must be demonstrated to be effective. This demonstration is typically done as part of summative usability testing. Through objective evidence, it must be shown that users will refer to, comprehend, and use the information for safety, such that it produces the expected reduction in risk.

18.5 SAMPLE RISK CONTROLS

Risk Controls should be a narrative to a specific action. Each Risk Control can be implemented via various means, such as System Requirements, operating procedures, and conformance to recognized standards. Some example Risk Controls that can be used as models are as follows:

- The device is designed to withstand up to 10 N of tensile pull-force
- The GUI is designed to minimize erroneous button pushes

- Standard of care is to take the product out of sterile pack with sterile gloves in the sterile field, in the OR
- Current leakage in the applied part is limited to 10 μA per IEC 60601-1

The Risk Control narratives in the Risk Assessment and Control Table (RACT) must be traced to their implementation means, and to verifications of implementation and effectiveness.

18.6 RISK CONTROLS AND SAFETY REQUIREMENTS

It is best to link Risk Controls to product requirements. This takes advantage of the formality with which requirements are tracked, change-controlled, traced, and verified. This formality also assures that Risk Controls will be implemented and that if any change to a requirement is contemplated, the impact of that change on the Risk Control will be evaluated.

When a requirement is linked to a Risk Control, it becomes a *safety requirement*. It is a good practice to tag safety requirements for ease of discernment by other parts of the product development team such as: R&D, Quality, Test Engineering, and Regulatory.

The connection between safety requirements and risk management is maintained via traceability. Fig. 25.1 depicts the connections between safety requirements and Risk Controls.

18.7 COMPLETENESS OF RISK CONTROLS

ISO 14971 [3,7] requires that the manufacturer ensure that the risks from all identified Hazardous Situations are considered and that the results of this activity are recorded in the Risk Management File. To meet this requirement, it is important that the risk analysis process is faithfully followed. This would ensure that all the relevant Hazards and Hazardous Situations are identified. Recording all the relevant Hazards and Hazardous Situations in the RACT, and logging their respective Risk Controls, residual risks, and relevant Harms would demonstrate compliance with this requirement of ISO 14971 [3,7].

CHAPTER 19

Risk Evaluation

Abstract

ISO 14971 requires the manufacturer to determine if the residual risks posed by the medical device are acceptable using the criteria defined in the risk management plan. Three methods of risk evaluation are presented: qualitative, semi-quantitative, and quantitative.

Keywords: Risk evaluation; qualitative; semi-quantitative; quantitative

ISO 14971 [3,7] requires the manufacturer to determine if the residual risks posed by the medical device are acceptable when compared against the criteria defined in the risk management plan (RMP). Although preliminary risk evaluations can be done before the completion of design, what matters is the final evaluation of the residual risks of the medical device, after all the Risk Controls have been implemented.

ISO 14971 [3] requires that the manufacturer evaluate the acceptability of residual risk for:

- each Hazard
- each Hazardous Situation
- overall

The Standard [3,7] does not prescribe a method for risk evaluation and allows the manufacturer to choose an appropriate method. Nor does the Standard specify whether the same or different criteria are to be used for the evaluation of individual residual risks, and the overall residual risks. The choice is left up to the manufacturer.

Depending on which method of risk estimation is chosen: qualitative, semiquantitative, or quantitative, the evaluation method varies.

19.1 APPLICATION OF RISK ACCEPTANCE CRITERIA

For some of the Hazards that a medical device presents, it may be possible to identify an applicable product safety standard that offers specific requirements for safety. ISO 14971 [3,7] considers compliance with such a standard an indication of acceptability of risk for that particular Hazard. According to Ref. [22], harmonized standards should be sought first. If no harmonized standard is available, then other national or international recognized standards or publications should be considered.

Safety Risk Management for Medical Devices
DOI: https://doi.org/10.1016/B978-0-12-813098-8.00019-2

© 2018 Elsevier Ltd.
All rights reserved.

Comparison of the risk of a medical device with the state-of-the-art is another way to establish acceptable risk. State-of-the-art in the context of ISO 14971 [3] is defined as "what is currently and generally accepted as good practice" [3,7]. "State-of-the-art does not necessarily mean the most technologically advanced solution" [3,7]. If it can be shown that for the same benefit, the risk of a medical device is less than or equal to the state-of-the-art, then it follows that the risk of the medical device is acceptable. Conversely, if you can show that for the same level of risk, your product delivers more benefit, you could argue that the risk of the medical device is acceptable.

How to determine state-of-the-art—One way is to review the literature to find applicable published scientific papers on the same or similar devices. Sound inclusion and exclusion criteria are very important as they may be subjected to scrutiny at a later date.

If the medical device is a new iteration of an existing approved device that is produced by the same manufacturer, evaluation of the relevant field data in the manufacturer's files could be another means of collecting data on state-of-the-art.

It should be noted that "good practice" in the definition of state-of-the-art does not limit the scope to medical devices. If, e.g., a pharmaceutical option is the currently and generally accepted best treatment for a disease, then the risk of that pharmaceutical option sets the bar against which to compare the risks of a medical device that provides a therapeutic option for the same disease.

The BXM method uses quantitative methods for risk evaluation. As such, the risk acceptability criteria also need to be quantitative. State-of-the-art risk for each Harm severity class is defined as the level of risk that the public accepts for therapy by comparable medical devices. For example, let's say a comparable medical device which has been in the market for some time, has a reported history of five cases of permanent, life-threatening injury to patients on a base of 10,000 patient-years of use. Since the predicate device is approved and continues to be marketed as a sufficiently safe device, the risk level of 0.05% per patient-year can be construed as an acceptable risk limit for the severity class of Critical, for that type of device/therapy.

In some cases, such as when a new and novel device is produced, or when a significant iteration on an existing device is made to deliver novel new therapies, state-of-the-art data may not be available. In such cases, risk acceptance criteria are derived from evaluating clinical study data, and formal benefit–risk analyses. See Chapter 23, Benefit–Risk Analysis, for more information on how to perform a benefit–risk analysis. It is noteworthy to realize that in such novel situations, the device under analysis could establish the state-of-the-art.

Stakeholder concerns—"It is well established that the perception of risk often differs from empirically determined risk estimates. Therefore the perception of risk from a wide cross section of stakeholders should be taken into account when deciding what risk is acceptable. To meet the expectations of public opinion, it might be necessary to give additional weighting to some risks" [3]. For example, people fear flying more than driving a car. As such, a large weighting factor has been put upon flight safety. The FAA sets the acceptable maximum probability of a catastrophic failure per flight hour for commercial aircraft to 10^{-9}, defined as extremely improbable. To put this in perspective, if an aircraft flies 10 hours per day, every day, it would take 273,973 years to complete 10^9 hours of flying.

As a practical means of measuring the public opinion about the tolerance of risk, it may be acceptable to presume that the concerns of the identified stakeholders reflect the values of society and that these concerns have been taken into account when the manufacturer referenced a reasonable set of stakeholders.

19.2 RISK EVALUATION FOR QUALITATIVE METHOD

In the qualitative method, risks are stratified into relative rankings from high to low. In the example offered in Section 17.1, three ranks of high (*red*), medium (*yellow*), and low (*green*) were devised.

For Hazards or Hazardous Situations for which applicable international or national standards can be found, determine residual risk acceptance by evaluation of conformance to those standards. For other Hazards and Hazardous Situations compare the individual residual risks and the overall residual risk to state-of-the-art. A method for establishing the state-of-the-art using risk-profiles was described in Section 11.3.1.

In the absence of applicable standards or state-of-the-art information, you may need to rely on subjective expert opinion to qualitatively judge the acceptability of risks of your device in consideration of the benefits that it offers.

19.3 RISK EVALUATION FOR SEMIQUANTITATIVE METHOD

Similar to the qualitative method, for the semiquantitative method, strive to use compliance with applicable international or national product safety standards as indication of risk acceptance. And, where applicable standards are not available, compare to the state-of-the-art, or refer to benefit—risk analyses.

In the semiquantitative method, we can numerically compare the probability of occurrence of Harms against the state-of-the-art. However, the severity is not easily comparable. That is, you could say a particular Harm, e.g., infection, has a severity of Serious and could happen with a probability 10^{-4}. That fills one cell in the risk matrix (see Section 17.2 and Fig. 17.2). But the state-of-the-art rating of severity may not match your severity rating. That is, the manufacturer(s) of the products that inform the state-of-the-art may have different definitions for the severity classifications than you do.

In the semiquantitative method, the comparison with the state-of-the-art is easier than in the qualitative method, but still equivocal.

19.4 RISK EVALUATION FOR QUANTITATIVE METHOD

The BXM method uses quantitative methods for risk estimation and evaluation. In the quantitative method as in the other methods, where possible, use compliance with applicable international or national product safety standards as indications of risk acceptance. Where applicable standards are not available, compare to the state-of-the-art, or refer to benefit–risk analyses.

In the quantitative method, comparison with the state-of-the-art is relatively straight forward. Compute the residual risk for a Hazard, a Hazardous Situation, or overall in the five severity classes of Catastrophic, Critical, Serious, Minor, and Negligible. See Section 17.3 for a detailed explanation on how to compute risk. If the computed residual risk for any severity class is larger than the acceptable risk limits that are spelled out in the RMP, then the risk is not acceptable. Otherwise, the risk is acceptable.

Just because a risk computes to be unacceptable, it doesn't mean the medical device cannot be released to the market. If benefit–risk analysis shows that the benefits of the device outweigh its risks, it may still be possible to get regulatory approval for the medical device.

CHAPTER 20

Risk Assessment and Control Table

Abstract

Every risk management process typically brings the results of the hazards analyses, risk estimations, and risk evaluation into a table. This is usually a large table and is called by many names. Examples: risk matrix, risk table, risk chart, product risk assessment, and risk analysis chart. The RACT tells the story of how each System Hazard manifests itself, what causes it, how exposure to that Hazard happens, and what Harms could ensue. The RACT also captures the Risk Control option analysis. That is, it shows what Risk Control options were considered and implemented. Additionally, the RACT computes the risks, both individual risks and overall System risk.

Keywords: Risk Assessment and Control Table; Risk Control option analysis; risk evaluation; RACT; residual risks

Every risk management process typically brings the results of the hazards analyses, risk estimations, and risk evaluation into a table. This is usually a large table and is called by many names. Examples: risk matrix, risk table, risk chart, product risk assessment, and risk analysis chart. The BXM method calls this table the Risk Assessment and Control Table (RACT). You can find a template for the RACT in Appendix B—Templates.

The RACT is a tool of risk management that integrates the results of all the analyses performed at lower levels of the System and enables the computation of residual risks for the System. The RACT tells the story of how each System Hazard manifests itself, what causes it, how exposure to that Hazard happens, and what Harms could ensue. The RACT also captures the Risk Control option analysis. That is, it shows what Risk Control options were considered and implemented. Additionally, the RACT computes the risks, both individual risks and overall System risk. The RACT can also evaluate the risks by comparing the computed risks against the risk acceptability criteria.

The BXM RACT template that is provided in Appendix B - Templates, does not show a pre-/post risk. Only the *final* risks of the medical device are shown. The reasons for this are:

1. The final risks of the device are what matter to the patient, and what is used to evaluate the benefit−risk ratio of the device.
2. Manageability of the RACT size. RACTs are typically huge tables and span multiple pages. Page management and reading of a larger RACT is more difficult.

DOI: https://doi.org/10.1016/B978-0-12-813098-8.00020-9
© 2018 Elsevier Ltd.
All rights reserved.

If you desire to track the pre/post risks, the modification of the RACT template is easy. Simply replicate columns P1 through *Negl-Risk*, and label them "pre." Alternatively, as the RACT is a living document and continues to evolve, you can save an earlier version of the RACT as the "pre-risk," and use the last version of the RACT as the "post-risk."

20.1 RISK ASSESSMENT AND CONTROL TABLE WORKFLOW

In the following sections the workflow for RACT is described. The workflow description corresponds to the template that is provided in Appendix B—Templates.

20.1.1 Examine the Clinical Hazards List

Start by citing all the *applicable* Hazards from the Clinical Hazards List (CHL) in the Hazard column. Entries in the CHL that are not applicable need to be rationalized in the Hazard Analysis Report. If new Hazards are identified in the underlying analyses, include them in the RACT and initiate an update to the CHL.

20.1.2 Capture End-Effects with Safety Impact

Examine the System Design Failure Modes and Effects Analysis (DFMEA), the System Process Failure Modes and Effects Analysis (for integral Systems) (PFMEA), and the System Use-Misuse Failure Modes and Effects Analysis (UMFMEA). Import the End-Effects that have a Safety Impact of Y. These End-Effects go into the Hazards column of the RACT.

20.1.3 Populate the Initial Cause and Sequence of Events columns

For the entries that are collected from the underlying Failure Modes and Effects Analyses (FMEAs), capture the Initial Cause of Hazard and Sequence of Events from the FMEAs and populate the RACT. It is possible that you may identify additional initial Causes and sequences of events. Add those to the RACT.

20.1.4 Populate Hazardous Situations column

Define the Hazardous Situations. A Hazardous Situation is where persons/property/environment are exposed to Hazards such that Harm can be experienced. For example, an implantable defibrillator battery may experience high current discharge and overheating. This creates a Hazardous Situation if the device is implanted in a patient.

20.1.5 Revisit the Preliminary Hazard Analysis

Revisit the Preliminary Hazard Analysis (PHA). If there is any relevant information in the PHA that is missing from the RACT, capture it.

20.1.6 Populate the P1 column

P1 is the probability of occurrence of the Hazardous Situation. Occ ratings from the System FMEAs can be inputs to the derivation of P1 values. For example, in the case of implantable devices, exposure is automatic. $P(\text{Hazard}) = P(\text{Hazardous Situation})$. But in other cases, where exposure to the Hazard is probable, $P1 = P(\text{Hazard}) \times P(\text{exposure})$.

20.1.7 Populate the Risk Controls columns

Risk Controls column is divided into three subcolumns: Safe by Design, Protective Measures, and Information for Safety. This is to support Risk Control option analysis as required by ISO 14971, Section 6.2 [3,7]. In the Risk Controls columns write narratives that describe the strategies to control the risks of the Hazardous Situations. The Risk Controls could apply to the initiating Cause, or somewhere in the sequence of events, or by preventing exposure to Hazards. Capture relevant mitigations from the underlying FMEAs. At the discretion of the analyst, the safety Requirement references may be entered in the Risk Control columns.

20.1.8 Populate the Harm column

Entries in the Harm column are taken from the Harms Assessment List (HAL). By design, all the potential Harms that the product can induce have been accounted for in the HAL. If a Hazardous Situation could precipitate multiple Harms, replicate the row and create a new row for each Harm.

20.1.9 Populate the P2 columns

P2 numbers are simple lookups from the HAL. They represent the probability of sustaining different severities of Harm from the same Hazardous Situation.

20.1.10 Compute risks

For each row, multiply P1 by the 5 P2 numbers to compute risk of Harm in five severity classes. This can be easily automated in a spreadsheet.

20.1.11 Risk evaluation

If quantitative risk acceptance criteria are defined in the Risk Management Plan, the risk evaluation could be automated by numerical comparison of the computed risks against the risk acceptance criteria.

20.2 INDIVIDUAL AND OVERALL RESIDUAL RISKS

In a quantitative risk management method, such as the BXM method, the per-Hazard, per-Hazardous Situation, and overall residual risks can be computed using Boolean algebra, and evaluated automatically.

For qualitative or semiquantitative methods, the manufacturer would need to create internal policies for the determination and evaluation of risks.

CHAPTER 21

On Testing

Abstract

Testing, in general is not a Risk Control. Rather, testing helps identify errors in design or implementation. Testing can also be used to build confidence in the design choices, and verify the implemented Risk Controls. One of the benefits of risk management is to provide a basis for risk-based sample-size determination for testing. This in turn can help steer resources toward the safety-critical aspects of the System.

Keywords: Testing; Risk Control; sample size; confidence; reliability

Testing can show the presence of errors, but not their absence.
Edsger Dijkstra, computer scientist (May 11, 1930–2002)

Testing, in general is not a Risk Control. Rather, testing helps identify errors in design or implementation. Testing can also be used to build confidence in the design choices, and verify the implemented Risk Controls. An exception to this is when testing is used to eliminate faulty products from reaching the market. For example, quality control (QC) testing at the end of the manufacturing process may be able to detect and reject products that have a safety defect. This type of testing does not eliminate the Hazard, but does eliminate exposure to the Hazard, and therefore it influences the P_1.

21.1 TYPES OF TESTING

During the course of product development, many types of testing are employed. Below, a few examples that are relevant to Risk Management are cited.

- Concept testing
  This type of testing is done to gain insights into the physics of operation of the medical device. It can demonstrate the feasibility of a concept, and give insights into Failure Modes. This type of testing can support assertions of failure Occ ratings. Concept testing is sometimes confused with mitigations, or Risk Controls. It is not the same. Testing doesn't reduce the rate of occurrence of a failure—it only supports the assertion of Occ rating.

Safety Risk Management for Medical Devices
DOI: https://doi.org/10.1016/B978-0-12-813098-8.00021-0

© 2018 Elsevier Ltd.
All rights reserved.

- Verification testing
  In the domain of Risk Management, verification testing is performed to demonstrate that Risk Controls are implemented as intended.
- Validation testing
  In the domain of Risk Management, validation testing is performed to demonstrate that the Risk Controls are effective in reducing risk.
- QC testing
  This type of testing is relevant to the manufacturing process. It can be used as a Risk Control to prevent defective parts/products from escaping to the field.

21.2 RISK-BASED SAMPLE SIZE SELECTION

It is wise to choose the rigor of your testing based on the safety risks related to the subject of the test. Choice of the sample size is an indication of the level of rigor. A higher sample size connotes higher confidence in the test results. This strategy can also have an economic benefit in that for low-risk test subjects, we can reduce the sample size and thereby save on product development costs.

The main question is: how many samples should be used for testing a safety-related requirement. Before we dive in to the methodology, it is beneficial to define two important terms: *Confidence*, and *Reliability*.

Confidence $(1 - \alpha)$—Probability that if the test passes, the requirement is met by at least $R\%$ of the population, where R is the *Reliability*.

Safety-requirements can be tested as variable or attribute. Attribute testing results are discrete, e.g., true/false or Pass/Fail, whereas variable test results are numerical on a scale. In the following sections, a strategy is offered on how to make risk-based decisions on sample sizes.

21.3 ATTRIBUTE TESTING

Requirements that are attribute-tested produce results that are discrete, usually binary. For example, the requirement: "The implantable device shall be able to withstand a 3T MRI environment without damage" would be attribute-tested. Multiple samples would be exposed to a 3T MRI environments, and then tested to see if they suffered any damage.

Using the BXM method, the following steps can be utilized to make a risk-based determination of sample sizes for attribute testing. The core idea is to identify the risk levels associated with the safety requirements and then determine the sample sizes commensurate with their risk levels. Safety requirements are those requirements that realize the Risk Controls.

Using the Risk Assessment and Control Table (RACT) as a source for risk determination, follow these steps:

Step 1. Examine the RACT to identify the Risk Controls which link to the safety requirement in question.

Step 2. Identify the Hazards that are linked to the Risk Controls which were identified in Step 1.

Step 3. For the aggregate of all the Hazards that were identified in Step 2, compute the residual risk in each Harm severity class. Use Boolean algebra's De Morgan's Theorem and Law of Identity.

Step 4. Within the aggregate identify the Harm severity class with the peak risk value.

Step 5. Use Table 21.1 to determine attribute-test sample-size that correlates to the severity class, which was identified in Step 4.

Note—The sample sizes in Table 21.1 are based on zero failures in all the tested samples.

Table 21.1 Confidence/Reliability (C/R) and attribute sample sizes

Catastrophic	Critical	Serious	Minor	Negligible
Conf./ Reliability 95/99	Conf./ Reliability 95/95	Conf./ Reliability 90/90	Conf./ Reliability 80/80	Conf./ Reliability 70/70
Attribute sample size 299	Attribute sample size 59	Attribute sample size 22	Attribute sample size 8	Attribute sample size 4

Note that the Confidence/Reliability numbers in Table 21.1 are merely suggestions. You should decide the numbers that are appropriate for your Quality Management System. The main point is to have a defensible and documented rationale for your decisions on sample size selection.

Table 21.2 shows sample sizes for other confidence and reliability combinations based on the binomial probability distribution. Assumption: zero failures in all the tested samples.

Table 21.2 Attribute-test sample sizes

Reliability lower bound (%)	95% Confidence	90% Confidence
70	$n = 9$	$n = 7$
80	$n = 14$	$n = 11$
85	$n = 19$	$n = 15$
90	$n = 29$	$n = 22$
95	$n = 59$	$n = 45$
96	$n = 74$	$n = 57$
97	$n = 99$	$n = 76$
98	$n = 149$	$n = 114$
99	$n = 299$	$n = 230$

21.4 VARIABLE TESTING

Variable testing of safety requirements produces numerical results along some scale. Example requirement: "Length of the connector shall be 3.00 ± 0.25 mm."

Using the same method that was described in Steps 1–4 of Section 21.3, determine the Harm severity class for the requirement in question. Identify the chosen confidence/reliability values. For variable test data, the statistical distribution of the data matters. Do a normality test on a preliminary dataset, e.g., a confirmation run. If the data distribution is normal, calculate the average and standard deviation of the data. Using a statistical software such as Minitab iteratively compute the minimum sample size needed to obtain a normal-distribution tolerance-interval that is narrow enough to fit within the specification. In the example of the connector above, that would be within ± 0.25 mm. If the computed sample size comes out to be very small, you can increase it to a larger number, e.g., 10 or 15.

If your data is not normal, you can perform a conversion of the data to make it normal, including the specification limits.

If your data indicates a different distribution, such as Weibull and Lognormal, still find the minimum sample size such that the tolerance interval falls within the specifications. You may need to consult a statistician.

Another method that is used is to correlate the risk level to process capability. For example:

For catastrophic risk class—CpK 2.0
For critical risk class—CpK 1.5
For serious risk class—CpK 1.0
For minor/negligible risk class—CpK 0.8

CHAPTER 22

Verification of Risk Controls

Abstract

Risk Controls must be verified for both implementation and effectiveness. Objective evidence is required to establish and support claims of verification.

Keywords: Risk Controls; verification; testing; implementation; effectiveness

Risk Controls must be verified for both implementation and effectiveness. Objective evidence is required to establish and support claims of verification.

The primary means of verification is testing. But sometimes visual inspections are used to verify the implementation of a Risk Control. For example, if the Risk Control is the installation of a red light, visual inspection for the presence of the red light is the easiest way to verify implementation. Caution should be exercised with respect to reliance on visual inspections, particularly if the visual inspection is done by one tester over a long period of time. It has been shown that after 2 ½ hours of continuous visual inspection, the propensity for error increases significantly. Also, visual inspection is subjective.

Be alert to optical illusions. There are many examples of optical illusions such as the images below. In the left image, (A) appears to be smaller than (B), while they are the same length. In the right image the horizontal lines appear to not be parallel, when in fact they are parallel.

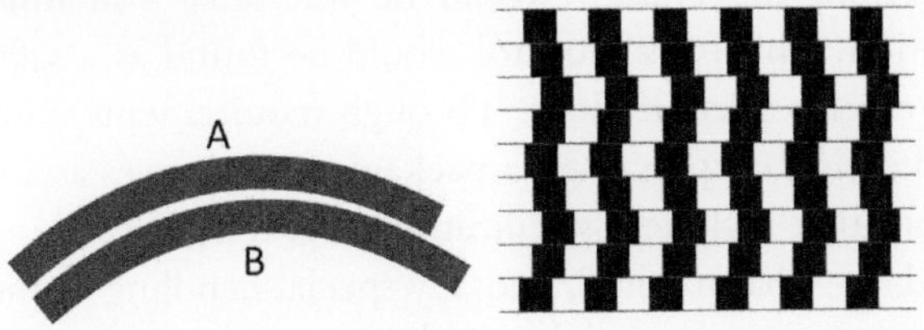

When doing testing, clear objectives and expectations are important. Human beings need to know what is important and needs their attention. Often people are given ambiguous objectives and expected to focus on many things. In such circumstances, it is easy for a tester to miss the things that they should notice.

Safety Risk Management for Medical Devices
DOI: https://doi.org/10.1016/B978-0-12-813098-8.00022-2

© 2018 Elsevier Ltd.
All rights reserved.

Whether attribute or variable testing is chosen, it is best if the selected sample size for testing is based on the safety risks. Sections 21.2–21.4 provide details about risk-based sample size selection.

22.1 VERIFICATION OF IMPLEMENTATION

Verification of implementation means providing objective evidence that the Risk Control is implemented. It can be in the form of test, inspection, demonstration, or analysis. Ensure that the subject of the test represents the final design. Otherwise, you'll have to repeat your test when the final form is available.

Include in your test protocols, any preconditioning, configuration, and adaptation if applicable. The protocol must include pass/fail criteria and the test method must be validated. The test report should include:

- The date on which the test was executed
- The names of individuals who performed the test
- The test protocol and methods used
- Any equipment/tooling used in the test including calibration and qualification evidence. If software is used as part of test equipment, refer to evidence of validation of that software
- Collected data or reference to it
- The test results and pass/fail declarations
- Any additional observations or deviations from the test protocol

Be sure to retain all test subjects, raw data, and anything else that you would need in case of an audit, to be able to defend or replicate your test results.

Let's take the example of an implantable medical device. A potential Harm from this device is infection; the Hazard would be microbial contamination. The Risk Control is sterilization. This Risk Control would be found as a safety requirement in the System Requirements Specifications. Through requirements flow-down and traceability, we can find design outputs, e.g., a packaging that serves as a sterile barrier, and process specifications that include a sterilization step. The IFU also mentions that the product is sterilized, by what method, and any special handling requirements.

To verify the implementation of this Risk Control, we'd look for design outputs such as drawing that show the product is protected by a sterile barrier. This would be by inspection. We'd look for test data that shows the sterile barrier meets its requirements. We'd also examine the manufacturing process design and identify the sterilization process step. We'd look for evidence of validation of the manufacturing process. Additionally, we'd inspect the IFU to find mentions of the fact that the product is sterile, by what method and any special handling requirements. This only verifies that the Risk Control is implemented. For a discussion on verification of effectiveness of risk controls, see Section 22.2.

22.2 VERIFICATION OF EFFECTIVENESS

Verification of effectiveness means providing objective evidence that the Risk Control is effective at reducing risk. Many methods can be used to establish that a Risk Control is effective in reducing risk. For example:

- *Usability testing*—Summative tests with real or surrogate users to show that the Risk Control measure reduces the risk. For example, if in a formative test using an earlier user interface design, it was found that half of the testers were making a particular mistake, and in the summative test we show that with the final design only 10% of the testers make the same mistake, we can conclude that the Risk Control was effective in reducing the likelihood of the Hazard, and thus risk.
- *Clinical study*—These are formal controlled studies performed on humans, typically to establish physiological effects. For example, let's say an older version of a pacemaker had a 10% rate of skin erosion in patients, and we have implemented a new design. A clinical study shows that the new design results in less than 5% of the test participants experiencing skin erosion. This would verify the effectiveness of the new design in reducing the risk of skin erosion.
- *Preclinical study*—These are formal studies performed on animals, or cadavers. They are used to test properties such as biocompatibility, biostability, toxicity, and efficacy. For example, let's say a new coating material is believed to reduce the rate of infection of an implantable device. In a preclinical study two samples of a device, one with the new coating and one without may be implanted in each animal. If under the same conditions, the samples with the new coating show a lower rate of infection, this would verify the effectiveness of the new coating in reducing the risk of infection.
- *Analysis and simulation*—It may be possible to verify the effectiveness of a Risk Control by analytical and simulation means. For example, if a new algorithm in implantable defibrillators is promised to be more effective in detection of ventricular fibrillation (VF), we could load the old and new algorithms in two copies of the same defibrillator and play recordings of a set of ECGs of VF in both devices. If the device with the new algorithm detects more VF episodes than the old algorithm, this verifies the effectiveness of the new algorithm.
- *Leverage verification*—In some cases the simple functionality of a Risk Control is proof of its effectiveness in controlling risks. For example, if a fuse is used to prevent high current flow, it can be shown in verification testing that the fuse cuts the current flow when current reaches a certain level. That verifies that the risk of electric shock is reduced.

One easy way to establish effectiveness of Risk Controls is by conformance to harmonized standards. For example, IEC 60601-1 [8] gives allowable electrical leakage current values, and even how to perform the tests. If you can demonstrate conformance to IEC 60601-1 with respect to electrical leakage current, you can presume that your Risk Controls are effective in reducing the risk of electrical shock. Another good source is IEC 60601-1-8 [19]. This Standard provides guidance on alarm systems in medical electrical systems. It provides requirements for alarms of high, medium, and low priorities, and says what color they should be, whether they should be flashing or constantly on, and if flashing at what frequency and duty cycle. This is very beneficial to the manufacturers, because it is very difficult to provide objective evidence that, e.g., a particular frequency of flashing red light is effective in reducing risk. But conforming to IEC 60601-1-8 [19] can be used as evidence of effectiveness of the alarm.

It is important that the test samples, test environment, and testers are representative of real life for the medical device. For example, if a medical device is intended for use by the surgical staff in an operating theatre, then the summative test should create a simulation of the ambiance of an operating theatre and utilize testers who are either members of a surgical team, or a good facsimile thereof. For home-use products, usability testing should use people who are representative of the intended users of the product. For example, if the product is intended for home-use by elderly people in France, then the testers should include elderly French people. This is because cultural and language variation from market to market makes a difference on how the users of the product perceive and understand the user interface and instructions that are provided.

If information for safety is used as a Risk Control, in accordance with IEC 62366-1 [14], that information needs to be tested on representative users to ensure that the information for safety is perceivable and understandable to the user, and it is actually effective in reducing risk. This means, e.g., if the user profile is elderly people with poor eye sights, the font and contrast of the information should be suitable for the user-profile group.

CHAPTER 23

Benefit–Risk Analysis

Abstract

ISO 14971 requires the manufacturers to establish that the benefits of the medical device outweigh its risks. To determine whether benefits of a medical device outweigh its risks, we must be able to answer two questions: 1) What are the potential benefits? and 2) What are the potential risks? Because the risks and benefits are typically not of the same units, it is often not possible to objectively balance the benefits against the risks.

Keywords: Benefit/risk ratio; benefit–risk analysis

Medical therapies are not static. Even if we do a benefit–risk analysis (BRA) now and show that the benefits of a device outweigh its risks, the same conclusion may not hold true later. Because as time goes on, new therapies are developed, culture and habits change, and benefit and risk perceptions evolve. For example, 60 years ago smoking was not only tolerated, there were even advertisement touting its health benefits. A simple internet search on vintage advertisements on the health benefits of smoking would show several cringe-worthy samples. Slogans such as "More doctors smoke Camels than any other cigarette," or "As your dentist, I would recommend Viceroys" may seem ridiculous today. But 60 years ago, they seemed perfectly reasonable.

Even political changes could affect health policy, reimbursements, and therefore alter people's aversion to risk.

BRA is a requirement of ISO 14971 for balancing the benefits of a device against the risks that it presents. To determine whether benefits of a medical device outweigh its risks, we must be able to answer two questions:

1. What are the potential benefits?
2. What are the potential risks?

Because the risks and benefits are typically not of the same units, it is often not possible to objectively balance the benefits against the risks. In other words, it is like comparing apples with oranges. Ultimately, a subjective decision must be made to answer the question:

Is the patient willing to accept the potential risk of the device for the potential benefit that it offers?

Safety Risk Management for Medical Devices
DOI: https://doi.org/10.1016/B978-0-12-813098-8.00023-4

© 2018 Elsevier Ltd.
All rights reserved.

Benefits of a device are presumed under the condition that the device works reliably, as intended. Risks and benefits are not the same for everyone. The same medical device could provide varying degrees of benefit to different patients and pose different levels of risks. Many factors are responsible for this, e.g., variability of the physiology and circumstances of a patient, environmental variability, and variability in manufacturing the medical device itself.

The FDA has published three guidance documents to help with the determination of benefit–risk balance for medical devices:

- Factors to Consider When Making Benefit-Risk Determinations in Medical Device Premarket Approval and De Novo Classifications [29]
- Benefit-Risk Factors to Consider When Determining Substantial Equivalence in Premarket Notifications [510(k)] with Different Technological Characteristics [30]
- Factors to Consider When Making Benefit-Risk Determinations for Medical Device Investigational Device Exemptions (IDEs) [31]

In general, the FDA expects the manufacturer to provide a "reasonable assurance of safety and effectiveness" by "weighing any probable benefit to health from the use of the device against any probable risk of injury or illness from such use."

Establishment of benefits is typically done by clinical evaluations. These are formal, well-planned investigations with defined end-points to show with statistical validity, the benefits of a medical device. See Annex X of the MDD [4] for further guidance on Clinical Evaluations. Sometimes nonclinical evaluations can be used to establish benefit. For example, animal- and cell-based testing, usability testing, and computer modeling and simulations.

Factors that are considered to evaluate the benefits of a device include:

- *Type of benefit*—Medical devices provide a variety of benefits. For example:
  - improving quality of life
  - reducing the probability of death
  - aiding improvement of patient function
  - reducing the probability of loss of function
  - providing relief from symptoms
- *Magnitude of the benefit*—This is typically measured against a scale according to specific criteria. For example, for Parkinson's disease the patients are asked to walk for 6 minutes. The longer the distance walked, the higher the magnitude of the benefit.
- *Probability of the benefit*—Not all patients receive the intended benefit. Sometimes it is possible to predict which patients are more likely to benefit from a given therapy. For example, a cancer therapy may be more effective if

the disease is diagnosed earlier in the course of the disease. It may be that the benefit is experienced by only a small subset of patients in the target population. Or conversely the benefit may be experienced by a large population.

- *Duration of the benefit*—Some treatments are curatives, while others may be only short term. The longer the benefit persists, the more the benefit would be valued when balancing the benefit against risk.

Patients are an important source of input on the determination of benefit–risk balance. They have an intimate and personal understanding of what it is like to live with a disease and how a device would impact their daily life. Ultimately, it is the patient who takes the risk for the promise of the benefit of a device.

The estimation of residual risks of a medical device is the domain of risk management, which is covered in detail in this book. Residual risk estimation can be quantitative or qualitative. Both clinical and nonclinical methods can be used to confirm estimations of risks, and the effectiveness of implemented Risk Controls.

When risk estimation is subjective and qualitative, the following factors are considered:

- *Severity, type, likelihood, and duration of Harms*—These are Harms that are caused by the use of the device. Risk management estimates the severity and likelihood of the Harms. Some Harms last a short while, such as a skin cut. But other Harms may last a long while. For example, brain damage would last a lifetime.
- *Procedure-related Harms*—Some Harms are not caused by the device, but are incidental. For example, implanting a medical device requires surgery, and surgery always presents risks of Harm.

When judging the risk versus the benefits of a device, additional factors that are considered by the FDA are:

- *Quality of the clinical data*—Poorly designed and executed clinical investigations could render the results of the investigation unreliable and weaken the argument of the dominance of benefit over risks.
- *The disease characteristics*—Factors considered are: whether the disease is degenerative and untreated worsens over time, or, is it stable, or does it get better over time; these make a difference in the judgment of the benefit–risk balance.
- *Patient risk tolerance*—The degree of risk tolerance varies greatly among patients and for different diseases. Patients with very severe diseases where the risk of dying is imminent may tolerate higher risks. For chronic diseases where patients learn to adapt to the presence and management of the disease, risk tolerance may be lower.

Risk tolerance is influenced by many factors including:
- The values of the society
- The population at risk (children, adults, etc.)
- Available alternative therapies
- The nature of the disease (acute, chronic, degenerative, etc.)
- Trust in the manufacturer

- *Availability of alternative therapies*—For some diseases alternative therapies may be available. For example, pharmaceuticals and gene therapy. If a device provides a benefit, though small, for a disease for which no alternative therapy exists, it may still get approved.

Sometimes a medical device shows greater risk than a comparable device on the market, but it may be approved to add to the portfolio of available treatments for patients. A higher-risk device may also get approved, if it shows greater efficacy for a subpopulation of the patients or under specific conditions, e.g., hot tropical environments.

There are certain circumstances where demonstration of acceptability of benefit to risk ratio is not difficult. Consider Fig. 23.1:

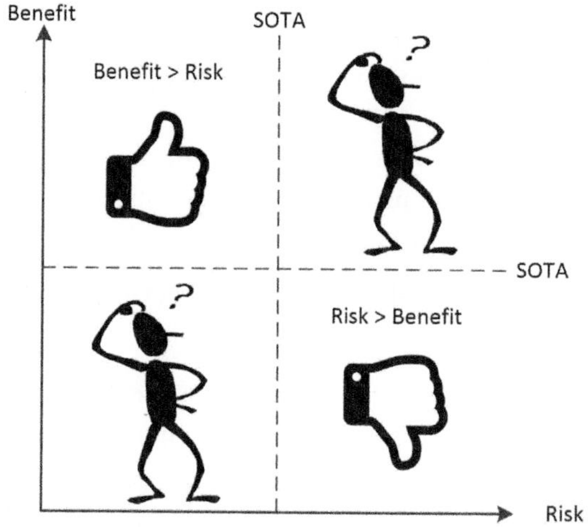

Figure 23.1 Risk—benefit comparison.

In this figure, the State-of-the-Art (SOTA) risks and benefits are depicted by *dotted lines*. SOTA represents a medical device that has been approved and in use already. Its benefits are known, and its risks have been accepted by the market. SOTA may be your own company's previous device, or a competitor's device. If the new device which is under analysis delivers more benefit for the same or lower risk, then the benefit–risk ratio is improved and clearly benefit outweighs the risk. This is represented by the upper left quadrant of Fig. 23.1. If on the other hand, the new device delivers less benefit for the same or more risk, then clearly the benefit does not outweigh the risk. This is represented by the lower right quadrant of Fig. 23.1. These are the clear cases. The upper-right and lower-left quadrants are not so clear and require a deeper analysis.

The degree to which different patients benefit from a medical device varies. BRA should be done on the overall target population. It may be that the benefit to risk ratio is acceptable for part of a target population and not for the entire population. For example, imagine a medical device that provides more benefit at a lower risk for a younger population than an older population. It may be determined that only patients under a cutoff age should be indicated for the use of the medical device.

23.1 BENEFIT–RISK ANALYSIS IN CLINICAL EVALUATIONS

Clinical evaluations or studies are a special case in the risk management of medical devices. In a clinical evaluation, the subject medical device may not be approved for commercial release. It may not have any benefits that the manufacture could claim. And, the clinical study may present safety risks to the study participants. Still the benefit/risk profile is expected to be evaluated and shown to be acceptable for the intended study target group.

For clinical studies, according to MEDDEV 2.7/1 [32], it is expected that the risks associated with the performance of the clinical study are minimized and acceptable when weighed against the benefits to the patient and are compatible with a high level of protection of health and safety.

With respect to the acceptability of benefit–risk ratio in clinical evaluations MEDDEV 2.7/1 [32] advises to evaluate the clinical data on benefits and risks of the device under evaluation when compared with the current SOTA in the corresponding medical field.

Per MEDDEV 2.7/1 [32] for clinical investigations, typically the following items are considered:

- clinical background
  - information on the clinical condition to be treated, managed, or diagnosed
  - prevalence of the condition
  - natural course of the condition
- other devices, medical alternatives available to the target population, including evidence of clinical performance and safety
  - historical treatments
  - medical options available to the target population (including conservative, surgical, and medicinal)
  - existing devices, benchmark devices

When making a decision on the acceptability of risks for the potential benefits, the deficiencies in current therapies should be identified from a critical and comprehensive review of relevant published literature to ascertain whether the investigative device addresses a significant gap in the currently available healthcare. If there is no significant gap, then the investigative device must show an improved or at least equivalent benefit/risk profile compared to existing products or therapies.

CHAPTER 24

Production and Postproduction Monitoring

Abstract

ISO 14971 requires that manufacturers collect and evaluate information about the medical device or similar devices, in the production and postproduction phases. The US CFR, title 21, part 822 has similar requirements. Patient/user safety is enhanced by active surveillance of production and postproduction information about marketed products. The manufacturer also receives significant benefit from the surveillance, namely, the opportunity to quickly identify and rectify product/process defects. This in turn leads into reduced customer complaints, reduced field corrective actions, improved reputation and customer loyalty, which means higher sales. And if that's not enough motivation, failure to perform product surveillance could result in substantial fines, criminal prosecutions, seizure of product, and closure of the business.

Keywords: Postmarket risk management; production and postproduction monitoring; surveillance; CAPA; HHA; risk management file review

Clause 9 of ISO 14971 [3,7] requires that manufacturers collect and evaluate information about the medical device or similar devices, in the production and postproduction phases. Similarly, under section 822 of CFR 21 the FDA requires the manufacturers and distributors of medical devices to perform postmarket tracking and reporting of device malfunctions, serious injuries, or deaths.

The term "postmarket" is used by the FDA in the United States. ISO 14971 [3,7] uses the terms "production" and "postproduction." These terms are roughly equivalent. Fig. 24.1 shows the contrast between these terms.

In the rest of this chapter we'll use the term "postmarket" interchangeably with "production and postproduction."

Production information is collected during the manufacturing process, and postproduction information is collected on all the phases of lifecycle after the product has been manufactured. The most significant postproduction lifecycle phase is product-use where majority of the feedback is generated.

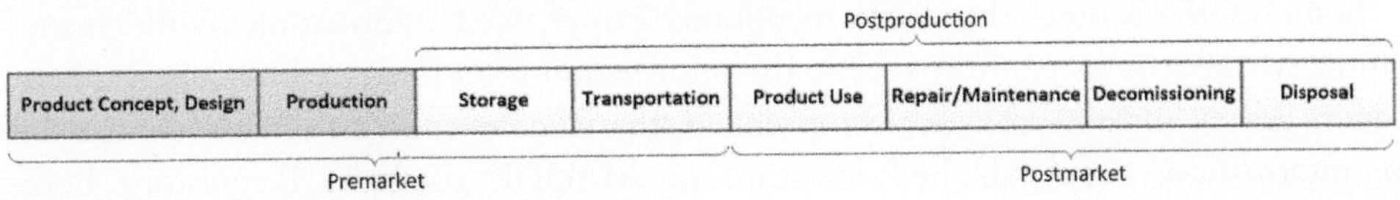

Figure 24.1 Postmarket versus postproduction.

DOI: https://doi.org/10.1016/B978-0-12-813098-8.00024-6

© 2018 Elsevier Ltd.
All rights reserved.

Listening systems should be established and tuned in to various sources of information. Examples of potential sources of information are as follows:

- manufacturing/production
- product service/repair
- returned product investigations
- customer service
- field service/customer visits
- clinical studies, e.g., postmarket clinical follow-ups
- customer complaints
- limited market releases
- product demonstrations
- databases such as MAUDE or EUDAMED
- legal department
- regulatory agencies, competent authorities (e.g., Field Safety Notices, Safety Alerts)
- suppliers
- distributors
- published literature on your device or similar competitive devices
- social media

Surveillance should include the applicable standards, such as ISO 14971. Changes to safety related standards may require revisions to your risk management procedures and risk management files.

The frequency of monitoring and trending of collected information should be established and documented. It is advisable that the frequency of data monitoring be based on the risk of the device. The higher the risk, the more frequent is the monitoring. For example, data monitoring may be executed monthly, quarterly, or yearly.

It is important to define the criteria for assessing the collected data. For example, you may trend the monitored data using established statistical methods and set up trigger criteria when trended data exceed predetermined thresholds.

One of the most important sources of customer feedback is complaints. Monitoring of this source of information is critical to the proper tracking and management of product field performance. In addition to providing insights to potential safety risks, complaint monitoring could provide valuable information on customer experience and product performance. This information could benefit product design teams to improve product design for better performance, safety, and customer satisfaction.

Some of the sources that were mentioned earlier, feed information to the manufacturer. Manufacturers passively receive the information from them and react to the information. An example is customer complaints. Other sources must be proactively searched for information, e.g., published literature or MAUDE database. Regulatory bodies increasingly look for more proactive postmarket surveillance of medical devices.

Risk management is applied to the entire product lifecycle, but it is very different in the preproduction versus production and postproduction. In the preproduction phase, risk management is a predictive tool. It forecasts what Hazards will manifest and what risks the medical device will present. After the product is produced and released for use, risk management is a retrospective tool, intended to protect the public from discovered risks.

Production monitoring aims to detect and prevent release of nonconforming products whose safety characteristics may have been adversely affected. Process data monitoring and trending, and quality control are some of the proactive ways in which production of nonconforming products is prevented, or their release avoided. If production monitoring indicates that nonconforming products with the potential to cause safety risks have escaped, then postmarket risk management activities should be initiated (see Section 24.1 for more details).

Postproduction monitoring is intended to evaluate the performance of the product in the field. A data monitoring plan should be established that stipulates sources of information and frequency of data collection and trending. Trending of data may show that a particular signal has been increasing and may have exceeded its threshold. Even though no patient Harm may have happened, this may indicate the potential for a future adverse event. Such a signal may necessitate the creation of a Corrective and Preventive Actions (CAPA) to initiate a root-cause analysis, determine preventive and corrective actions, and plan for verification of effectiveness of said actions. Sources that should be included in postproduction monitoring are:

- nonconformance reports
- CAPAs with potential for safety risk to patients
- complaint trending signals
- signals from returned product investigation, and service
- reports of new Hazards in competitor products
- published information
- postmarket clinical follow-ups
- surveillance registries
- reports of injuries and death that are attributable to the product

If postproduction monitoring indicates reported or potential Harm from released products, postmarket risk management activities should be initiated (see Section 24.1 for more details).

Postmarket-detected issues that have a potential safety impact must be addressed within certain time limits (e.g., 30 days) to limit the exposure of patients.

Sometimes root-cause analysis could identify a supplier of the manufacturer as the source of a Hazard. In that case, CAPA would reflect back to the supplier. The CAPA may also impact the incoming inspection procedures at the manufacturer.

24.1 POSTMARKET RISK MANAGEMENT

As stated earlier, preproduction risk management is a predictive tool, used to forecast future Hazards and risks, while postmarket risk management is a reactive endeavor, intended to contain and limit Harm from devices that have been released to the market.

Production and postproduction monitoring are the means by which to detect whether a released product has harmed patients or has the potential for causing undue harm. The result of postmarket risk management may be to hold manufactured product from being released, or issue corrective actions, including product recalls. In cases of imminent potential Harm to patients, it may be necessary to issue immediate information (Customer Letter) before new Risk Controls are developed and verified.

Postmarket risk management involves understanding of: the Hazard in question, the risk of potential Harm, size and magnitude of the patients' vulnerability, and countermeasures that should be taken until the product itself is addressed for corrective measures.

A common tool that is used in this vein is the Health Hazard Assessment (HHA). In the United Stated, 21 CFR part 7 requires the conduct of an evaluation of the health hazard (actual or potential) presented by a product being recalled or considered for recall. In Europe, MDD [4] Article 10 compels similar action. Elements of an HHA are:

- Identify the Hazard (actual or potential).
- Identify the related Harm(s), both immediate and long term.
- Identify the population at risk and whether any subpopulation is at greater risk.
- Describe the mechanism of occurrence of the Hazard, and exposure to the Hazard.
- Is the Hazard in question happening when the product is used under labeled conditions?
- Is the Hazard in question previously predicted?
- Is the Hazard in question manifesting under normal or fault conditions?
- Identify conditions that would exacerbate or mitigate the risk.
- Identify the degree to which the Hazard is recognizable by the patient/user and the feasibility of countermeasures.
- What is the likelihood of future additional risks due to the Hazard in question?
- Balance the risk of corrective action versus the risk of not taking corrective action. Example: an implanted device may pose a greater risk than previously predicted. But the risk of explant surgery could be greater than the risk of leaving the device in.
- Conclusion and recommended actions.

Any decision for action or inaction should be approved by appropriate personnel, e.g., medical safety officer, and the decisions and approvals should be documented.

Postmarket risk management may determine measures to control the discovered risks. These Risk Controls may be in the form of design changes, labeling, or training. Naturally, the new Risk Controls must be verified for effectiveness. The risk management file may need to be updated. The manufacturer may also issue advisory/field safety notices.

For medical devices currently installed and used in the market, the Risk Control measures may be different from those applied to devices that are currently in production. For example, for products in the field, Risk Control measures may include sending information to doctors or patients, removing product from the field and providing replacement product. For products that are in production, Risk Control measures may include identification and collection of product, and rework or discard of product.

If a recall or field product update/revision is decided, speed of action will be material to the effectiveness of risk reduction activities.

24.2 FREQUENCY OF RISK MANAGEMENT FILE REVIEW

As postmarket surveillance continues to bring in new information about the product field performance, the risk management file should be periodically reviewed to determine need for update. The frequency of risk management file review depends on:

- the risks of the device
- the novelty of the device
- the duration of time that the device has been in the market

For a new, novel, and high-risk device do more frequent reviews; perhaps as much as monthly or bimonthly. The longer the device has been on the market and the more that has been learned about it, the review frequency can be reduced, e.g., every year or every other year. For low-risk, old-technology devices, the review frequency could be even lower, e.g., every 4–5 years.

Whatever frequency the manufacturer chooses should be recorded in the Risk Management Plan. Document the rationales for your decision on review frequency.

24.3 FEEDBACK TO PREPRODUCTION RISK MANAGEMENT

There is a connection between the postmarket and preproduction risk management. Knowledge gained from product performance after it is manufactured and released must be fed back to preproduction risk management in a drive to add veracity to the preproduction predictions and estimations of the risks of a medical device. The types of information that postmarket risk management can provide to preproduction risk management are:

- Is there evidence of new Hazards that were previously not foreseen?
- Are the P_1 and P_2 estimations still valid?

- If qualitative methods are used, are the estimates of risks still valid?
- Are there reports of misuse which were not foreseen in the original risk management process?
- Are the Risk Controls proving to be effective in reducing/maintaining risk levels?
- Is there any evidence that the actual market experience of risk acceptability for the received benefit has changed?

One of the key benefits of production and postproduction monitoring is the derivation of P_1 and P_2 numbers from field data. P_1 and P_2 estimates that are based on actual field data are far more credible than expert opinion. However, the key to successful mining of field data is thoughtful, well-planned and well-executed collection, coding and cataloging of the field data.

P_1 is the probability of occurrence of a Hazardous Situation. P_1 is prevalence. P_1 has units. The units of P_1 are dependent on the application and are determined by the manufacturer. For example, for an insulin pump a sensible unit could be patient-hours of operation. For sphygmomanometer (blood pressure monitor) a sensible unit could be number of uses. To measure P_1, a certain time interval needs to be selected.

A hypothetical example: 50,000 units of insulin pump model X have been in operation between Jan 1, 2017 and Jun 30, 2018. During this period, the average length of service for model X has been 4000 hours. In the same time period, there have been 250 reported and confirmed cases of over-infusion.

$$P_1 = 250/(50{,}000 \times 4000) = 1.25 \times 10^{-6} \text{ per patient-hour}$$

This information should be fed back to preproduction risk management so that if the prediction of P_1 for over-infusion was different from 1.25×10^{-6} it can be updated.

P_2 is the probability of sustaining Harm, given that the Hazardous Situation has already occurred. In the BXM method, risk is computed in five severity classes, as defined in Table 26.1.

P_2 is outcome-based. To derive P_2 from filed data, we need to know after the Hazardous Situation happened, what was the outcome for the patient. For example, in the insulin pump example earlier, we would ask, after the over-infusion event what was the Harm and what was the outcome to the patient? A potential Harm of insulin over-infusion is hypoglycemia. Potential answers to the above question could be:

- Patient died (catastrophic)
- Patient became unconscious and suffered brain damage (critical)
- Patient fainted and was taken to the emergency room, but has recovered now (serious)
- Patient felt a little light headed but ate a piece of candy and was fine (minor)
- Patient reported feeling a little strange, but it passed (negligible)

Each of the above answers would be counted as one instance in each of the five severity classes. Let's imagine that in a dataset of 97 hypoglycemia events we have the following counts:

- Catastrophic: 0
- Critical: 2
- Serious: 20
- Minor: 25
- Negligible: 50

P_2 is agnostic of the Hazardous Situation. That means we want to know, for all reported cases of hypoglycemia regardless of the Hazardous Situation that caused it, what were the outcomes. So, if two different Hazardous Situations could cause hypoglycemia, we aggregate all the hypoglycemia cases together.

Continuing with our example, the total cases of hypoglycemia were: 97. Therefore for the Harm of hypoglycemia P_2 numbers are:

- P_2(Catastrophic): 0/97 = 0%
- P_2(Critical): 2/97 = 2.1%
- P_2(Serious): 20/97 = 20.6%
- P_2(Minor): 25/97 = 25.8%
- P_2(Negligible): 50/97 = 51.5%

To ensure effective feedback loops from postmarket to preproduction risk management, it is advisable that the responsibility for maintaining the risk management file be defined and assigned to specific staff.

24.4 BENEFITS OF POSTMARKET SURVEILLANCE

Clearly patient/user safety is enhanced by active surveillance of production and postproduction information about marketed products. But the manufacturer also receives significant benefit from the surveillance, namely, the opportunity to quickly identify and rectify product/process defects. This in turn leads into reduced customer complaints, reduced field corrective actions, improved reputation and customer loyalty, which means higher sales. And if that's not enough motivation, failure to perform product surveillance could result in substantial fines, criminal prosecutions, seizure of product, and closure of the business.

Another benefit of postmarket surveillance is about the clinical Hazards list (CHL) (see Section 11.5). The CHL is an invaluable tool of risk management. The CHL is claimed to be complete, at any given time. The basis of that claim is that the CHL is a living document, containing the best available knowledge at any given time. And, if any new Hazards are discovered, they are added to the CHL. Without postmarket surveillance data, we cannot make the claim of completeness.

CHAPTER 25

Traceability

Abstract

ISO 14971 requires manufacturers to provide traceability for each Hazard to its risk analysis, risk evaluation, Risk Controls, residual risk evaluation, verification of implementation, and effectiveness of Risk Controls. This chapter examines methods and strategies for capturing and documenting traceability for medical devices.

Keywords: Traceability; software; Risk Management Report

ISO 14971 [3] requires manufacturers to provide traceability for each Hazard to its risk analysis, risk evaluation, Risk Controls, residual risk evaluation, verification of implementation, and effectiveness of Risk Controls. For medical devices that include software, IEC 62304 [9] requires that traceability between System requirements, software requirements, software system test, and Risk Control measure implemented in software must be made.

Traceability is an invaluable tool to ensure completeness in risk management. Without traceability, it is possible to miss Hazards, fail to control their risks, or fail to verify their Risk Controls. Fig. 25.1 depicts a model for traceability for risk management.

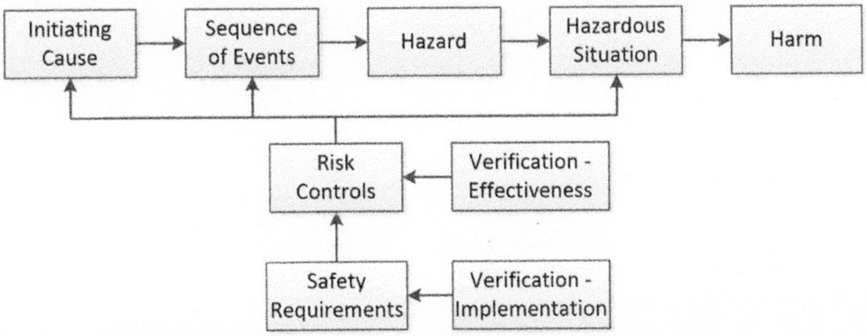

Figure 25.1 Traceability model.

You can capture your traceability in any form that is convenient. Examples include spreadsheets, databases, or requirements management systems.

Traceability between Hazards and risk analysis can be captured in the Risk Assessment and Control Table (RACT). That is, the description of initiating event

Safety Risk Management for Medical Devices
DOI: https://doi.org/10.1016/B978-0-12-813098-8.00025-8
© 2018 Elsevier Ltd.
All rights reserved.

and the subsequent sequence of events that lead into the Hazard is captured in the RACT. Similarly, the traceability between each Hazard and the evaluation of the risks associated with that Hazard can be captured in the RACT.

The assessment of acceptability of the residual risks must also be documented. In the BXM method this traceability is also captured in the RACT.

As design, implementation, testing, and risk management activities continue, it is crucial that the integrity of traceability links be maintained. Without this diligence, it is easy for the links to become inaccurate. The use of automation tools can help with this endeavor. Many requirements management tools enable creation of entities, e.g., requirements, Risk Controls, or test cases, and to link such entities within the tool. This is very convenient for up-to-the-minute viewing and reporting of link maps. Moreover, the tools can flag links as suspicious, if one or the other end of the link is modified. The engineers can then review the contents of both ends of a link and verify whether the link is still valid.

For devices of even moderate complexity the traceability analysis report would be a large document. For this reason, it is not recommended that the traceability analysis report be included in the Risk Management Report (RMR), as it could make the RMR too large and unwieldy. Instead, it is better to include a summary of the traceability analysis report in the RMR, and make a reference to the traceability analysis report, as one of the elements of the Risk Management File (RMF).

CHAPTER 26

Risk Management for Clinical Investigations

Abstract

Clinical investigations are governed by ISO 14155, which addresses good clinical practice for the design, conduct, recording, and reporting of clinical investigations carried out on human subjects. ISO 14155 requires that prior to the execution of clinical trials, the risks associated with clinical trials be estimated in accordance with ISO 14971.

Keywords: Clinical investigations; risk management; clinical study

In this chapter, we will examine the requirements for risk management in clinical investigations. Clinical investigations are governed by ISO 14155 [33], which addresses good clinical practice for the design, conduct, recording, and reporting of clinical investigations carried out on human subjects (also called participants). ISO 14155 [33] defines Clinical Investigation as a "systematic investigation in one or more human subjects, undertaken to assess the safety or performance of a medical device."

Clinical studies are conducted to increase medical knowledge as to how a medical device performs in humans. Some examples of the reasons for conducting clinical studies:

- Evaluate the clinical benefits of a medical device in treatment of diseases, syndromes, or conditions in the target patient populations.
- Evaluate the risks associated with the use of a medical device both due to the medical device itself, and how it would be used in clinical settings.
- Confirm the predicted risks and identify any new Hazards associated with the use of the medical device.
- Gain insight into uncertainties about the performance of the medical device in vivo.
- Identify rare complications.
- Examine the performance of the medical device under long-term and widespread use.
- Investigate particular features regarding clinical utility.
- Assess cost/benefit or health outcomes in support of reimbursements.

Safety Risk Management for Medical Devices
DOI: https://doi.org/10.1016/B978-0-12-813098-8.00026-X

© 2018 Elsevier Ltd.
All rights reserved.

26.1 TERMINOLOGY

In the following sections references are made to the following terms. It is important to understand the language of clinical investigations and be able to distinguish the terms.

Adverse Device Effect (ADE): "Adverse event related to the use of an investigational medical device" ([33] 3.1).

Adverse Event (AE): "Any untoward medical occurrence, unintended disease or injury, or untoward clinical signs (including abnormal laboratory findings) in subjects, users or other persons, whether or not related to the investigational medical device" ([33] 3.2).

Clinical Evaluation: A methodologically sound ongoing procedure to collect, appraise and analyze clinical data pertaining to a medical device, and to evaluate whether there is sufficient clinical evidence to confirm compliance with relevant essential requirements for safety and performance when using the device according to the manufacturer's Instructions for Use [32].

Clinical Investigation: Systematic investigation in one or more human subjects, undertaken to assess the safety or performance of a medical device ([33] 3.6).

Clinical Investigation Plan (CIP): Document that states the rationale, objectives, design and proposed analysis, methodology, monitoring, conduct and record-keeping of the clinical investigation ([33] 3.7).

Investigator's Brochure (IB): Compilation of the current clinical and non-clinical information on the investigational medical device(s), relevant to the clinical investigation ([33] 3.25).

Serious Adverse Device Effect (SADE): Adverse device effect that has resulted in any of the consequences characteristic of a serious adverse event ([33] 3.36).

Serious Adverse Event (SAE): Adverse event that

a. led to death,
b. led to serious deterioration in the health of the subject, that either resulted in
  1. a life-threatening illness or injury, or
  2. a permanent impairment of a body structure or a body function, or
  3. in-patient or prolonged hospitalization, or
  4. medical or surgical intervention to prevent life-threatening illness or injury or permanent impairment to a body structure or a body function,
c. led to fetal distress, fetal death or a congenital abnormality or birth defect ([33] 3.37)

Ref. [33] also notes that the terms "Clinical trial" or "clinical study" are synonymous with "clinical investigation."

26.2 CLINICAL STUDIES

Before we discuss the requirements of risk management for clinical studies, it is important to understand the different types of clinical studies.

There are many factors that determine the type of clinical study:

- Premarket, postmarket
- Exploratory, confirmatory, or observational
- Interventional, noninterventional

In the earliest stages of medical device development, it may be necessary to evaluate the merits and limitations of a new device, prove a concept, or test a new indication for an existing device. These are premarket, exploratory, or feasibility studies performed on a small number of participants and require risk management.

If the exploratory studies produce good results, a confirmatory pivotal clinical investigation can be performed to collect data on the safety and efficacy of the device on a larger group of participants. Pivotal clinical investigations also require risk management.

After the device is approved for market release, postmarket confirmatory clinical investigations can be performed on the device in order to collect data on the clinical performance, safety, and efficacy of the device. Risk management may be necessary depending on the objectives of the CIP.

Another type of clinical investigation is the observational postmarket study, where the device is used within its labeled indication. This type of study collects data on large groups of patients to evaluate specified outcomes on patient populations, and serves scientific, clinical, reimbursement, or policy purposes. Often these are registry studies where the decision to use the medical device is clearly separated from the decision to include the subject in the clinical study. Observational postmarket studies do not introduce any additional risk on the study participants and risk management is not required prior to the start of the study.

Ref. [33] notes that the term "noninterventional" is synonymous with "observational," and that "postmarket clinical investigation" is synonymous with "postmarket clinical follow up."

26.3 MAPPING OF RISK MANAGEMENT TERMINOLOGIES

ISO 14155 [33] defines SAE as an event that: "

a. led to death,
b. led to serious deterioration in the health of the subject, users or other persons defined as an adverse event resulting in one or more of the following:
   1. a life-threatening illness or injury, or
   2. a permanent impairment of a body structure or a body function, or
   3. in-patient or prolonged hospitalization, or
   4. medical or surgical intervention to <u>prevent</u> life-threatening illness or injury or permanent impairment to a body structure or a body function,
c. led to fetal distress, fetal death or a congenital abnormality or birth defect"

Table 26.1 Definitions of severity based on ISO 14971 table D.3

Severity class	Definition
Catastrophic	Death
Critical	Permanent impairment or life-threatening injury
Serious	Injury or impairment that requires professional medical intervention
Minor	Temporary injury or impairment that does not require professional medical intervention
Negligible	Inconvenience or temporary discomfort

A commonly used set of definitions for Harm severities, based on [3,7] table D.3 is presented in Table 26.1:

ISO 14155 [33] requires that the risks associated with the clinical investigations be estimated in accordance with ISO 14971 [3,7]. This necessitates the mapping of terminologies among the two standards. It could be surmised that SAE maps to Catastrophic, Critical, and Serious severities of Harm, and therefore, Minor and Negligible classes of Harm severity would NOT be SAE.

As it is intended that Clinical Investigations provide feedback to risk management regarding the risks of the medical device, a higher resolution classification of SAEs is needed to facilitate proper feedback to Risk Management. The CIP would be a good place to capture the higher resolution classification of AEs and SAEs during the Clinical Investigations.

26.4 RISK MANAGEMENT REQUIREMENTS

ISO 14155 [33] requires that prior to the execution of clinical trials, the risks associated with the use of the investigational device be estimated in accordance with ISO 14971 [3,7].

The participants in a clinical investigation are faced with two types of risk:

- Risks associated with the medical device
- Risks associated with the design and conduct of the clinical study

The risks associated with the medical device are the main subject of this book and are covered extensively. The risks associated with the conduct of the clinical study could involve the clinical study design, methods of data collection, data processing, clinical setting, personnel performance, etc.

Risk management must estimate and balance the *combined* risks against the potential benefits to the participants. The same principles that are applied to the management of the device risks are applied to the design and conduct of the clinical investigation. The clinical study risks must be analyzed, estimated, controlled, and evaluated.

When analyzing risks, consider the risks not only to the participants, but also to the clinicians, investigators, and other persons.

Benefit–risk analysis for clinical studies is different from the benefit–risk analysis of commercially released devices in that the point of the study is to demonstrate the benefit. As such, risk management considers the *expected* and *potential* benefits of the device in the benefit–risk analysis.

As part of the risk management of the clinical study, it is expected that a thorough review of published and available unpublished literature be done to uncover any known risks that could be relevant to the clinical study.

For a clinical investigation, risks are controlled over two horizons: before an AE, and after the AE. Before the AE, Risk Controls aim to prevent the Hazardous Situations from happening. After the AE, Risk Controls aim to limit the harm. Actions such as patient monitoring, AE reporting, or termination of the clinical investigation are designed to limit the harm to study participants. The study sponsor should design and implement appropriate training for the clinical investigators to ensure that proper data collection, processing, recognition, and escalation activities are performed. The extent and scope of this training should be based on the severity of the risks, as communicated from risk management.

Sometimes clinical studies that intend to investigate a new indication for an existing device use an approved device off label. The additional device-risk which is introduced from the off-label use is counterbalanced by the close clinical monitoring and safety protocols of the clinical investigation.

Clinical investigations are themselves a component of the risk management process in that they provide evidence of the benefits, which is used in the benefit–risk analysis.

At the end of the clinical investigation the risk data collected should be reviewed and fed back into the risk management process for confirmation or revision of the estimated risks. Also, if any new Hazards were identified during the study, they should be added to the risk analysis. The benefit–risk analysis should be revisited as well for confirmation or revision.

Any knowledge gained about the potential Hazards and risks of the medical device should be captured and communicated to the risk management process for use in future analyses of the risks of the subject device. Also, knowledge gained from the performance of the clinical investigation and risks that were manifested to the participants, users, or other persons should be captured and used in future clinical studies.

26.5 RISK DOCUMENTATION REQUIREMENTS

Risk management makes contributions to the required documents of clinical investigations. Table 26.2 outlines the contribution of risk management to elements of the clinical investigation documentation.

Table 26.2 Risk management input to clinical documentation

Clinical investigation document	Risk management contribution
Clinical Investigation Plan (CIP)	Summary of risk analysis Anticipated adverse device effects Identification of residual risks Risk Controls Benefit–risk analysis summary
Investigator's Brochure (IB)	Summary of risk analysis Anticipated adverse device effects Identification of residual risks Anticipated risks, contraindications, warnings, etc. Benefit–risk analysis summary Results of risk assessment
Clinical Investigation Report (CIR)	Benefit–risk analysis summary Adverse events Adverse device effects
Informed Consent Form (ICF)	Anticipated adverse device effects

CHAPTER 27

Risk Management for Legacy Devices

Abstract

For established manufacturers, it's likely that they have products that have been in the market for a long time. Perhaps even before the existence of ISO 14971. These products are termed Legacy Devices; defined as medical devices which were legally placed on the market and are still marketed today, but for which there is insufficient objective evidence that they are in compliance with the current version of the Standard. This chapter provides guidance on how to manage the risks of Legacy Devices.

Keywords: Risk management; legacy device

For established manufacturers, it's likely that they have products that have been in the market for a long time. Perhaps even before the existence of ISO 14971. These products are termed Legacy Devices; defined as medical devices which were legally placed on the market and are still marketed today, but for which there is insufficient objective evidence that they are in compliance with the current version of the Standard [3,7].

Since ISO 14971 is intended to be applied throughout the entire lifecycle, especially during the design phase, a retrospective application of the standard to an existing legacy device is not particularly valuable. However, an abbreviated application of the Standard [3,7] would be of value, particularly for postproduction risk management and maintenance of the risk management file.

The following steps may be performed as an alternative to performing clauses 4–6 of the Standard [3,7].

1. Ensure there is a risk management process in place that is compliant with clause 3.1 of the Standard [3,7].
2. Prepare a risk management plan for the legacy device in accordance with clause 3.4 of the Standard [3], [7].

   The scope of the risk management plan can be limited to: the creation of the risk management file, performance of production and postproduction risk management, and maintenance of the risk management file. The plan should define the actions and responsibility for field data collection and processing. If future versions of the device will be developed, the plan should lay out the appropriate activities for the risk management of the new device.

Safety Risk Management for Medical Devices
DOI: https://doi.org/10.1016/B978-0-12-813098-8.00027-1

© 2018 Elsevier Ltd.
All rights reserved.

3. Establish and maintain a risk management file per clause 3.5 of the Standard [3,7].
4. Identify the following for the Legacy Device:
   a. Intended use and clinical indication
   b. Intended patient population
   c. Intended user profile
   d. Intended use condition and environment
   e. Operating principle
   f. Characteristics related to safety
5. Considering item #4 above, identify the Hazards, Hazardous Situations, and potential Harms from the legacy device.
6. Identify the Risk Control measures that are already in place in the Legacy Device, and classify them as: Safe by Design, Protective Measures, or Information for Safety.
7. Create a traceability report among the Hazards, Hazardous Situations, Harms, and Risk Controls.
8. Create a risk management report to document the above activities.
9. If advances in technology and practices enable the manufacturer to feasibly further reduce the device risks, then in the future releases of the legacy device additional Risk Controls should be implemented and the risk management file updated, including a new benefit–risk analysis.

CHAPTER 28

Basic Safety and Essential Performance

Abstract

Basic Safety and Essential Performance are key concepts in medical device risk management. This chapter gives guidance on how to identify Basic Safety and Essential Performance, and how to distinguish them from each other.

Keywords: Basic Safety; Essential Performance; IEC 60601-1

IEC 60601-1 [8] is the Standard for nonimplantable medical electrical equipment. Ref. [8] defines two special terms:

- Basic Safety: "freedom from unacceptable risk directly caused by physical hazards when me equipment is used under normal condition and single fault condition" [8] 3.10.
- Essential Performance: "performance of a clinical function, other than that related to Basic Safety, where loss or degradation beyond the limits specified by the manufacturer results in an unacceptable risk.

  NOTE Essential Performance is most easily understood by considering whether its absence or degradation would result in an unacceptable risk" [8] 3.27.

Understanding, and the ability to identify the Basic Safety and Essential Performance of your medical device is essential to your ability to demonstrate compliance to [8].

28.1 HOW TO IDENTIFY BASIC SAFETY

Consider your medical device while it is *not* performing its clinical function. Analyze for all the relevant requirements of IEC 60601-1 [8] and determine if the device under normal and single fault conditions could pose a Hazard. This is the basis of identification of Basic Safety of your device.

28.2 HOW TO IDENTIFY ESSENTIAL PERFORMANCE

Essential performance is about clinical functions of the device. Follow these steps:

1. Make a list of all your clinical functions.

Safety Risk Management for Medical Devices
DOI: https://doi.org/10.1016/B978-0-12-813098-8.00028-3
© 2018 Elsevier Ltd.
All rights reserved.

2. Identify those functions whose failure or degradation could lead into a Harm. These are candidates for Essential Performance.
3. Use the Risk Assessment and Control Table to determine the risk levels for each candidate function from Step (2). If the residual risk for a clinical function is unacceptable, then that clinical function is part of the Essential Performance of the device.

Note that it is possible to have a medical device with no Essential Performance. What that means is that the device risks are controlled in such a way that the loss or degradation of any of its clinical functions do not result in unacceptable risk.

CHAPTER 29

Relationship Between ISO 14971 and Other Standards

Abstract

ISO 14971, the central Standard for medical device risk management, works in concert with many other related Standards. In this chapter we examine the relationships between ISO 14971 and IEC 60601-1, ISO 10993-1, IEC 62366, and ISO 14155.

Keywords: IEC 60601-1; ISO 10993-1; IEC 62366; ISO 14155

As mentioned in Chapter 5, Risk Management Standards, a number of safety-related Standards rely on ISO 14971 for determination of safety risks. In the following subsections, the relationships between ISO 14971 and IEC 60601-1, ISO 10993-1, IEC 62366, and ISO 14155 are described.

29.1 INTERACTION WITH IEC 60601-1

IEC 60601-1 [8] is related to general requirements for basic safety and essential performance of Medical Electrical equipment (ME), and is intended for the nonimplantable MEs.

IEC 60601-1 [8] makes a normative reference to ISO 14971 [3] and expects the performance of risk management per Ref. [3].

IEC 60601-1 [8] specifies certain requirements and acceptance criteria that facilitate the risk management process. For example, there are detailed requirements in Ref. [8] for protection against electric shock. Test equipment, measurement methods, and pass criteria are provided in the Standard [8]. Therefore if a medical device is compliant with IEC-60601-1 [8], it can claim that the risk of electric shock from the device is acceptable.

In other cases, e.g., with respect to the safety of emitted ultrasound energy from a medical device, Ref. [8] does not offer any means of testing for safety or acceptance criteria. Instead, it defers to the risk management process per ISO 14971 [3].

In some cases, the determination of whether certain aspects of the medical device design are subject to IEC 60601-1 [8] is left up to the risk management process. For example, "applied part" has a special meaning in Ref. [8]. It is the part of the device

Safety Risk Management for Medical Devices
DOI: https://doi.org/10.1016/B978-0-12-813098-8.00029-5

© 2018 Elsevier Ltd.
All rights reserved.

that comes into contact with the patient for the purpose of performing its clinical function. In some cases, a medical device may come into contact with the patient/user but not for the purpose of performing the device's clinical function. Such contact may be a source of Hazards. IEC 60601-1 [8] defers to ISO 14971 [3], the determination of how to treat the nonapplied parts of the device that come into contact with the patient/user.

The concept of Essential Performance was defined in Section 3.1 and elaborated in Chapter 28, Basic Safety and Essential Performance. IEC 60601-1 [8] expects the manufacturer to identify the Essential Performance of the medical device. There are many specialized test houses who perform IEC 60601-1 [8] compliance testing for manufacturers. As input to the test house, besides samples of the medical device, the test house expects the manufacturer to identify the Essential Performance of the device.

A point to keep in mind about using compliance with IEC 60601-1 [8] as the foundation of claim of risk acceptance, is that IEC 60601-1 [8] itself is not a Risk Control. The design features that enable passing the IEC 60601-1 [8] tests, are the Risk Controls. Therefore the 60601-1 [8] passing test results are the objective evidence that the Risk Controls are effective.

29.2 INTERACTION WITH ISO 10993-1

ISO 10993 is a series of Standards numbered ISO 10993-1 to ISO 10993-22. ISO 10993-1 [15] is about biological evaluation and testing of medical devices within a risk management process. ISO 10993-2 to ISO 1-003-22 are each dedicated to a special aspect of biocompatibility. The aim of this series of Standards is the protection of humans from biological risks due to the use of medical devices.

The range of biological responses that are considered as adverse, i.e., Hazards, is quite broad and complex, and vary from person to person. They include: cytotoxicity, sensitization, irritation, hemocompatibility, pyrogenicity, carcinogenicity, and genotoxicity.

ISO 10993-1 [15] does not intend to provide a rigid battery of tests, including pass/fail criteria that can be used by a manufacturer to demonstrate biocompatibility. Instead, the standard offers guidance on ways of determining what testing is needed with aim of achieving an acceptable level of risk for the benefits that the medical device offers.

ISO 10993-1 [15] considers biological risk management as part of the overall process of management of safety risks to humans in accordance to ISO 14971 [3,7]. Conduct of biological evaluations serves to meet the requirements of both ISO 14971 [3,7] and ISO 10993-1 [15].

Execution of biological evaluations requires planning. The biological evaluation plan is a part of the overall risk management plan. It's possible to combine the two plans under one cover. The biological evaluation plan must include arrangements for the identification, estimation, and evaluation of biological risks of the medical device. Additionally, the plan must include activities for the identification, review, and approval of: biological Risk Controls, residual biological risk, and disclosure of the residual biological risks.

One of the activities of biological risk analysis is the identification of biological hazards. This involves the consideration of the level of toxicity of utilized materials, and their route and duration of exposure. Sometimes the physical properties of materials play a role in the level of toxicity of the material. This includes factors such as surface roughness or porosity of the material. Chemical properties of materials can also introduce biological hazards. For example, dissimilar metals may create a galvanic and corrosive process, which creates a biological hazard.

As part of Hazard identification, potential contributors such as the choice of materials, additives, processing-aids, and catalysts should be considered. Also, downstream processes, e.g., welding, sterilization, degradation materials, and packaging can introduce Hazards.

Characterization of the biological hazards of a device includes identification of the type and duration of exposure to materials. For example, exposure to intact skin is very different from exposure to internal tissues such as blood or brain.

Risk estimation requires knowledge of the probability of exposure to the toxicant and the potential Harm that could ensue. Probability of exposure can be derived from the availability of the toxicant and the dose–response in the exposed tissue. Severity is related to the biological response and can be estimated based on published literature or animal studies.

Risk Controls aim to reduce the potential risks. Some examples of Risk Controls are:

- Design changes to eliminate toxic materials from the design or manufacturing process
- Changes to physical/geometric properties
- Reduction of exposure time
- Avoidance of the more hazardous exposure routes

Biocompatibility can also guide the device risk management by contraindication of patients who could be more vulnerable. For example, a product may be adequately safe for adults, but not safe for infants.

Evaluation of the biological risks of a medical device is part of the overall residual risk evaluation. Risk evaluation requires knowledge of risk acceptance criteria. The criteria for risk acceptability are established in the risk management plan, at the start of the design process.

Risk Controls must be verified for implementation and effectiveness. Verification of implementation can typically be achieved via the normal design verification testing. Biological evaluation could serve as verification of biological Risk Control effectiveness.

A great service from biological evaluation to risk management is providing data, which allows for more accurate risk estimation, instead of simply assuming the worst-case outcomes.

Biocompatibility testing can be lengthy and expensive. A sound biocompatibility strategy can provide the justification to waive certain tests. This strategy could, e.g., be based on toxicology information or relevant prior use of materials. This is not only efficient, but also ethical. In some cases, it may be possible to reduce/eliminate animal testing by substituting chemical and in vitro testing.

29.3 INTERACTION WITH IEC 62366

IEC 62366 [14] is the Usability Engineering Standard. There are two types of formal testing that are envisioned in this standard: Formative and Summative. Formative tests are performed iteratively during the design and development phase, with the intention to explore the effectiveness of the user interface design and to identify potential use-failures or misuses. Formative tests typically do not have formal acceptance criteria. Their goal is to guide the design of the user interface (UI) and achieve a level of quality that ensures the summative tests will be successful. From the risk management perspective, the focus of formative testing is on reducing the likelihood of use-failures that have the potential for creating Hazards.

Summative tests are performed at the end of user interface development, after formative tests are completed. The objective of summative tests is to produce objective evidence that the user interface contribution to safety risks is acceptable. Summative tests are qualitative investigations where observations can be reported in simple statistics, e.g., 8 out 10 testers succeeded. Summative tests have formal acceptance criteria and can be utilized as a means of verification of effectiveness of Risk Controls.

Risk Management can inform Usability Engineering vis-à-vis decisions on the performance of summative tests. One such contribution is on the identification of user interfaces that have, or have not, an impact on safety. For safety-related user interfaces, summative tests need to be performed. The choice of the number of participants in a summative test can be informed by Risk Management.

IEC 62366-2 [34] offers equation K.1 as:

$$R = 1 - (1 - P)^n$$

where

R is the cumulative probability of observing or detecting a usability problem
P is the probability of a single test participant having a usability problem
n is the number of test participants

Given the risk acceptance criteria and P2 numbers, the maximum probability of the Hazard and thereby probability of use-failure (P) can be derived. Additionally, there can be a "Company" policy that R, the probability of detecting a use-failure in a summative test must be e.g., $\geq 90\%$. With these two pieces of information, the number of participants can be derived. For example: Let's say max tolerable P is 10%, and R, per "Company" policy is 90%. The number of participants, n computes to 22. You can also use table K.1 in Ref. [34] for the same purpose.

Another input from Risk Management to Usability Engineering is the indication of areas of the UI that are not safety related, thereby enabling Usability Engineering to use expert reviews, rather than conduct new summative tests, for minor changes to those areas.

The contribution of Usability Engineering to Risk Management is both in the identification of use-failures which could result in Hazards, and in the performance of summative studies which can serve as verification of effectiveness of Risk Controls.

29.4 INTERACTION WITH ISO 14155

ISO 14155 [33] is about clinical investigation of medical devices in human subjects. ISO 14155 [33] makes a normative reference to ISO 14971 [3,7] and requires that prior to the conduct of clinical investigations, the risks associated with the use of medical devices in clinical investigations be estimated according to ISO 14971 [3,7].

ISO 14155 [33] states that the terms: clinical investigation, clinical trial, and clinical study are synonymous. As such, these terms are used interchangeably herein.

In a clinical investigation, risks can be viewed over two horizons: prestudy and poststudy. Before the study, risk management per ISO 14971 [3,7] serves to identify the Hazards to the study participants, estimate the risks, control the risks, and do benefit–risk analysis. In this capacity, risk management behaves as a predictive engineering tool. After the study has begun, risk management is reactive to Adverse Events and attempts to minimize any Harms to the study participants.

There is a bidirectional flow of information between risk management and clinical investigations. In fact, clinical investigations are a part of the risk management process.

Flow of information from risk management (ISO 14971) to clinical investigations (ISO 14155):

- Identification of Hazards; estimation of risks; control of risks
- Identification of potential Harms, their nature, severity, probability, and duration
- Information-for-safety, warnings, cautions
- Acceptability of benefit–risk profile
- Any special risk areas on which the clinical study should focus
- The risk management report

Flow of information from clinical investigations (ISO 14155) to risk management (ISO 14971):

- Confirmation of benefits
- Confirmation of risk estimates
- Verification of effectiveness of Risk Controls, user interface design, and Information For Use (IFUs)
- Confirmation of predicted Hazards and discovery of any new Hazards

There are also additional activities that provide information to both clinical investigations and risk management, namely:

- Structured scientific literature reviews to identify any reported known and potential Hazards, risks, performance issues, benefits, and claims. Literature reviews include same and similar devices, and also nonsimilar devices that have similar elements as the subject device.
- Identification and evaluation of alternatives available to patients.
- Review of complaints data.
- Review of Medical Device Regulations.
- Review of public databases, e.g., FDA MAUDE adverse event report database (United States), and EDUAMED (European).

CHAPTER 30

Risk Management Process Metrics

Abstract

Process metrics are interesting to many businesses, and the risk management process is no exception. But how does one measure the performance of a risk management process?

Keywords: Process metrics; risk management process

Process metrics are interesting to many businesses, and the risk management process is no exception. But how does one measure the performance of a risk management process?

If a project was successful, met its objectives, had no questions/findings from the Regulatory bodies, and the product is performing safely in the market, is that a sign that the risk management process was successful? Or, could it be just good luck? The problem with measuring risk management is that when risk management is successful, the product gets approved smoothly and no one gets injured. No drama at all!

So, how do we go about measuring the effectiveness and success of a risk management process? What do we measure? How do we measure it? And, if we were able to measure "it," what is considered good, and what's bad? What is the criteria for goodness?

There is currently no consensus on how to measure the goodness of a risk management process. But in the subsections below, we offer three options for consideration.

30.1 COMPARISON WITH HISTORICAL PROJECTS

If a company has produced and commercialized a significant number of medical products, and has collected data about each product on how smoothly the product was approved, and whether it was the subject of any field corrective actions, then the company could potentially create a benchmark from a composite of the performance of the previous products. With this benchmark, risk management process on a new product could be measured.

The problem with this method is that it is a lagging indicator. It can only indicate whether the risk management for a project met expectations, perhaps years after all is said and done.

Safety Risk Management for Medical Devices
DOI: https://doi.org/10.1016/B978-0-12-813098-8.00030-1

© 2018 Elsevier Ltd.
All rights reserved.

30.2 ISSUE DETECTION HISTORY

Similar to the method in Section 30.1, this method relies on historical performance and collection of data. But this method can provide a leading indicator on the performance of a risk management process.

During the course of risk management processes, issues with potentially adverse safety impact are identified and mitigated. These are usually design-related issues which had gone unnoticed by the product development team. Assuming constant maturity and performance by the design team, one can presume and expect a certain rate of detection of safety issues. If the rate of issue detection on a new project is significantly lower than the historical levels, one could construe that the current risk management process is not working as well as the past projects.

The problem with this method is that team performance is not constant, project complexities are not the same, and safety issue identification can be subjective. So, the conclusions derived are just conjectures.

30.3 SUBJECTIVE EVALUATION

In this method people who are involved in the project *and* experienced with risk management evaluate the project on several vectors. For example:

- the productivity of the working sessions and how well they are run
- the sense of confidence in the ability to identify Hazards and estimate risks
- the contribution of the risk management process to a sense of communication among the participating functional groups
- how well risk management identifies safety impact of proposed design changes

This method is also not precise, but can provide real-time feedback to management. And, even though it is subjective, it is not less valuable than the other two methods that are offered in this chapter.

CHAPTER 31

Risk Management and Product Development Process

Abstract

As technology advances, more sophisticated and more complex devices are produced to handle difficult medical conditions. Many of these devices employ hazardous and potentially lethal sources of energy such as gamma radiation and lasers, or dispense life supporting medicines into the patients. The rise in complexity of these medical devices brings about both increased benefits and increased safety risks. What safety strategies should a manufacturer adopt? How can risk management be a value-added activity to product development? Can risk management help identify Essential Design Outputs? This chapter explores answers to these questions.

Keywords: Safety strategy; essential design outputs; lifecycle; product development

As technology advances, more sophisticated and more complex devices are produced to handle difficult medical conditions. Many of these devices employ hazardous and potentially lethal sources of energy such as gamma radiation and lasers or dispense life supporting medicines into the patients. The rise in complexity of these medical devices brings about both increased benefits and increased safety risks.

The requirements of ISO 14971 [3,7] with respect to risk management apply to the entire lifecycle of a medical device. This includes the product development part of the lifecycle.

Considering product risks during the product development process, the safety features of the System can be developed as an integral part of the product development process and effectively integrated in the System architecture. This reduces the product development costs and can reduce the product development schedule as well.

With advance consideration to safety, several characteristics should be considered in the design. For example:

- The safety aspects of the System should be as simple as possible, with clearly understandable design and operation.
- Functionality of safety-critical parts of the System should be kept independent of the rest of the System, if possible.
- Interfaces to safety-critical parts of the System should be well defined.

Most medical Systems today are complex. It's likely that some latent design flaws with the potential for unintended behaviors would remain in the design. It is advisable

Safety Risk Management for Medical Devices
DOI: https://doi.org/10.1016/B978-0-12-813098-8.00031-3
© 2018 Elsevier Ltd.
All rights reserved.

to create functional and design boundaries between the safety-critical parts of the System and the rest of the System and create firewalls to limit the impact of the latent design defects.

In the course of regulatory review of the medical device submissions, the device design, particularly the safety aspects of the design become scrutinized. A well-architected System with clear and simple safety subsystems would be more easily reviewed, with fewer questions, and gets approved more quickly.

31.1 IDENTIFICATION OF ESSENTIAL DESIGN OUTPUTS

CFR Title 20, Part 820.30 [35] states that "Design output procedures shall contain or make reference to acceptance criteria and shall ensure that those design outputs that are essential for the proper functioning of the device are identified."

It can be stated that Essential Design Outputs (EDOs) are those design outputs, which are essential to the safety and efficacy of the medical device.

The term "design output," which is referenced in 21 CFR 820.30(d), can have different meanings. A design output is the result of design efforts at each phase, and also at the end of the total design effort. Therefore there can be intermediate and final design outputs.

The FDA defines *finished* design output as the basis for the Device Master Record (DMR); and defines the *total* finished design output of the device, as the device itself plus its packaging, labeling, and the DMR.

So, what are EDOs? Why do they matter? And, once we identify them, what are we supposed to do about them? Let's answer each question in turn.

1. EDOs are element of the actual finished design output whose loss or degradation would have an adverse impact on the safety or proper functioning of the device. For example, a particular dimension on a part, or a component itself could be EDOs.
2. EDOs matter because they are essential for the "proper functioning" of the device; meaning safe and effective functioning of the device.
3. Once EDOs are identified, certain policies should be exercised to provide higher confidence in the implementation and performance of those outputs. This could be in the form of increased process capability requirements, tighter QC inspections, etc.

So, what role does risk management play for EDOs? As stated earlier, EDOs are critical for the safe and effective functioning of the device. Risk management is concerned with safety, and can assist the product design process in the identification of EDOs. Fig. 31.1 presents a strategy for the identification of EDOs from a safety

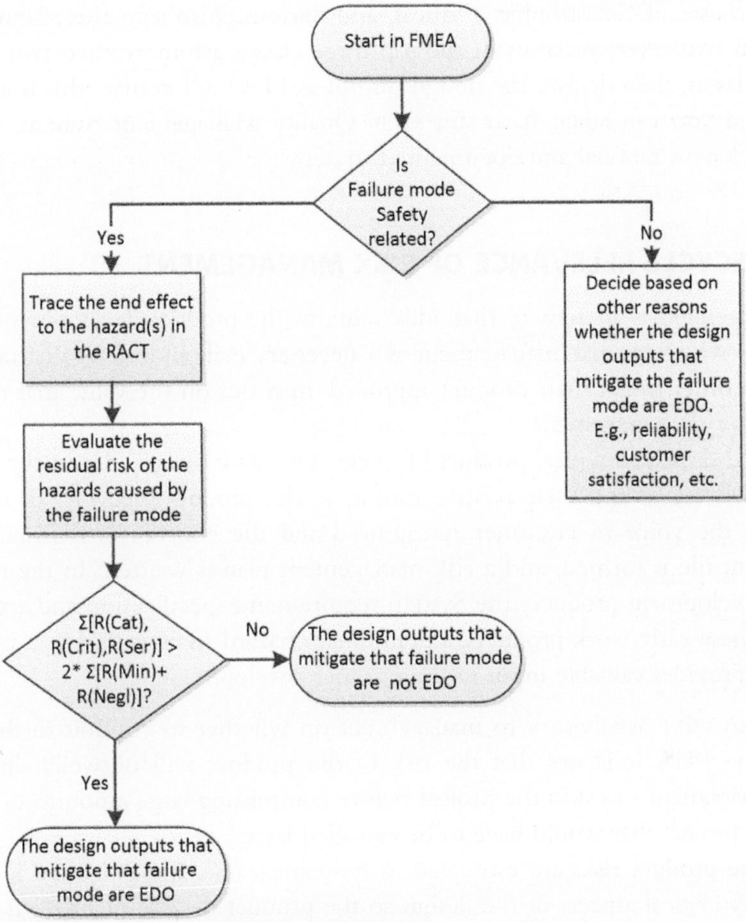

Figure 31.1 Identification of Essential Design Outputs.

perspective. This strategy is designed around the BXM method, but can be adapted for other methods as well.

The core concept is the use of Failure Modes and Effects Analyses (FMEAs) to help with the identification of EDOs. Not every Failure Mode has a safety impact. For nonsafety-related Failure Modes, other reasons could drive the decision as to whether a design output is EDO. For instance, reliability or customer satisfaction could cause the manufacturer to declare a design output as EDO. If the Failure Mode does have a safety impact, then we need to know the amount of risk it presents to the patient/user. A safety-related Failure Mode is necessarily connected to one or more Hazards of the System. Use the Risk Assessment and Control Table to determine the residual risk of the Hazard(s) that are the result of that Failure Mode. You should end up with 5 numbers, one for each severity class of Harm. Sum the residual risk for the top three classes of Catastrophic, Critical, and Serious. Also sum the residual risk for the bottom two severity classes. If the top three classes are more than twice the bottom two classes, then declare the design output as EDO. Of course, this is a suggested strategy and you can adapt it to suit your Quality Management System. The main point is to have a rational and documented strategy.

31.2 LIFECYCLE RELEVANCE OF RISK MANAGEMENT

Risk management is an activity that adds value to the product development lifecycle. Companies who view risk management as a necessary evil, and a box that needs to be checked in order to get their product approved, miss out on the value that they could derive from risk management.

Fig. 31.2 displays a typical product lifecycle. The *black triangles* depict the risk management deliverable and their relative timing to the product lifecycle. In the beginning, after the voice of customer is captured and the concept is released, the risk management file is formed, and a risk management plan is written. In the meantime, product development produces the System requirements specification and architecture. Based on these early work products, a preliminary hazard analysis (PHA) is performed. The PHA provides valuable input to the product development:

1. It can serve as advisory to management on whether to commit to the project. If the PHA indicates that the risk of the product will outweigh its benefits, management can stop the project before committing large amounts of resources to a project that would have to be canceled later.
2. If the product risks are estimated to be manageable, the PHA can identify the safety critical aspects of the design so the product development team can enter the design phase with knowledge of where to focus their resources. This serves to reduce waste and optimize resource usage in product development.

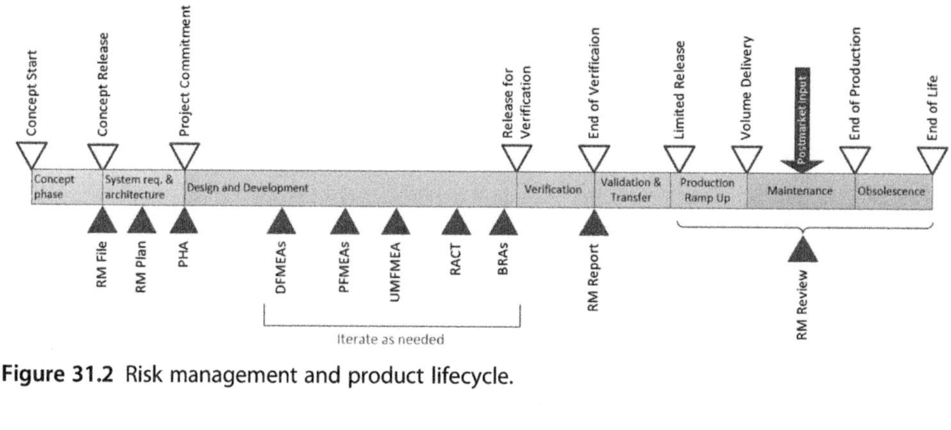

Figure 31.2 Risk management and product lifecycle.

After the approval of the project and commitment of resources, design and development begins in earnest. As designs of the product, the process, and the user interfaces become available, FMEAs begin. The FMEAs provide feedback to the design team to improve the design. Another benefit that the risk management process offers to the product development team is the estimation of risk, which not only enables the teams to make risk-based sample size determination for verification testing, but also alerts the design team to potential problem areas. The sooner the design team knows about the problem areas, the less costly it is to fix them.

After risk assessment is completed, benefit–risk analyses are performed to provide evidence that the benefits of the device outweigh its risks. This is a critical part of the regulatory submission, without which you cannot get approval for commercialization of your product.

After the release of the product, risk management continues to monitor the product in the production and postproduction phases. New knowledge gained about the product performance is fed back to the risk management process, and evaluated for potential updates and/or improvements to the prelaunch risk estimates.

CHAPTER 32

Axioms

Abstract

Axioms are self-evident truths upon which we build our knowledge and analysis.

Keywords: Axioms; system safety; safety vs. reliability

Ten axioms of medical device risk management that would be useful to keep in mind are provided below:

1. Safety is not a function, but an attribute
   Customers acquire medical devices for clinical benefits, and expect safety
2. Hazardous Situations can arise even when there are no faults, i.e., under normal operational conditions
3. Safety is an emergent property of the System
   Knowledge of safety of System components does not assure safety of the System
4. A Hazard cannot result in Harm until a sequence of events leads to a Hazardous Situation
5. Safety and reliability are not the same thing
   See Chapter 13, Safety Versus Reliability, for more details
6. Severity is a qualifier for Harm
7. Death is not a Harm
   Death is a potential consequence of a Harm
8. Risk Controls are targeted at risk reduction (severity | likelihood)
9. Software is never a Hazard; but can contribute to a Hazard
10. Highly reliable software is not necessarily safe

Safety Risk Management for Medical Devices
DOI: https://doi.org/10.1016/B978-0-12-813098-8.00032-5

© 2018 Elsevier Ltd.
All rights reserved.

CHAPTER 33

Special Topics

Abstract

In this chapter some special topics are covered that are not per se, part of the risk management process, but are of interest to practitioners of medical risk management.

Keywords: Personal liability; combination medical devices; complacency; Cassandras

In this chapter some special topics are covered that are not per se, part of the risk management process, but are of interest to practitioners of medical risk management.

33.1 THE CONUNDRUM

Human psychology tends to become myopic and lose sight of potential dangers, if there are no perceptible signs of danger. This is why most governments tend to *react* to disasters at a much higher cost, than *prevent* disasters at a lower cost.

Likewise, when risk management predicts high risk of an adverse event, if there have not been any reported occurrences of the adverse event, people tend to think "it hasn't happened before, therefore likely it will not happen in the future." This kind of thinking diminishes the level of attention that high-risk events deserve.

Successful risk management results in nothing bad happening. This may lead to a lack of appreciation by the outsiders, including management. A sense of complacency could set in, and attention and investment in risk management can diminish.

Sometimes even political motivations and aspiration can get involved. Careers tend to get advanced when a person steps up in a crisis and saves the day. It is thought that Winston Churchill said: "Never waste a good crisis." Heeding warning of risk management to avert an unprecedented crisis could appear as wasting money and resources on something that has not happened before. And if nothing happens, the precautionary actions don't receive credit. On the other hand, ignoring the warnings of risk management and taking a chance could result in the crisis, which could create the opportunity for a "hero" to step up and fight the fires. It doesn't matter whether the hero succeeds or not. Heroism will be rewarded.

DOI: https://doi.org/10.1016/B978-0-12-813098-8.00033-7
© 2018 Elsevier Ltd.
All rights reserved.

33.2 CASSANDRAS

Cassandra was a character in Greek mythology who could foresee future disasters but was cursed by the gods so that when she would warn people, no one would believe her. This is a term used to refer to people who warn of future disasters but are not believed.

Richard Clarke, the former counterterrorism adviser to US presidents Bill Clinton and George W. Bush, has written a book: "Warnings: Finding Cassandras to Stop Catastrophes" [36]. In this book, he talks about how Cassandras can/should be recognized, and how to benefit from their foresight while not being buffeted by too much fear.

In medical device risk management, we are required to analyze risks from both known and foreseeable Hazards. Cassandras tend to better see the foreseeable Hazards. The problem is that if something has never happened before, it may be difficult to persuade your organization to devote resources to it.

Clarke suggests not to be dismissive. Instead, take a surveillance and hedging strategy. What this means is to spend a small amount of resources and monitor the foreseen Hazard. Perhaps do some experimentation, investigate, research, and gather data. If the data supports the forecast, then devote more resources to mitigate the Hazard. Otherwise, you may be able to disprove the hypothesis. You don't have to make a final decision at once. It can be taken in steps.

33.3 PERSONAL LIABILITY

Hiding safety-related defects or falsifying test results are illegal and carry serious legal consequences, including personal liability on the part of the perpetrators. For example, the *Wall Street Journal* published a story in Aug 9, 1996 that reported the conviction of three C.R. Bard executives and sentencing them to prison-time for knowingly conspiring to hide potentially deadly flaws in a catheter model, and selling devices that had not been approved by the FDA. In another example, *The Telegraph* published a story on Dec 10, 2013 about Jean-Claude Mas, the founder of PIP breast implants, who was sentenced to 4 years in prison for the deliberate use of unapproved silicone gel in breast implants.

It should be noted that this does not mean that if a medical device causes injury, it is automatically concluded that people who were involved in the design and production of that device are criminally liable. It is understood that even if manufacturers follow sound practices for risk management, and do all they can to prevent injuries to people, some injuries are inevitable.

33.4 RISK MANAGEMENT FOR COMBINATION MEDICAL DEVICES

For combination medical devices, which combine device, drug, or biologics, a good strategy is to first do analysis of risks for each aspect of the combination device, i.e., device part, drug part, and biologic part, and then analyze for the risks due to the interaction of the parts, such as drug with device. Then bring it all together in the Risk Assessment and Control Table and determine the overall residual risk.

CHAPTER 34

Critical Thinking and Risk Management

Abstract

Critical thinking is an intellectually disciplined process of receiving information and analyzing it accurately and objectively, free from bias.

Keywords: Critical thinking; risk management; thinking errors; cognitive traps

Critical thinking is an intellectually disciplined process of receiving information and analyzing it accurately and objectively, free from bias. It is easier said than done.

We all think. But without the discipline of critical thinking, much of our thinking is biased, distorted, and inaccurate. Yet in our daily lives we make decisions, sometimes with dire consequences, based on mindless thinking. This book is not about critical thinking—that is a much larger subject. But in this chapter a few examples of critical thinking issues are provided just to make the reader aware of the potential impact of thinking errors on risk management.

Below some contributors to thinking errors are highlighted.

Incredulity—We often miss things if they don't fit our mental models and beliefs. Can you imagine a color that you have never seen before? It is not possible, because you need to first have a mental model of the color, before you can imagine it. If a phenomenon that you believe cannot happen happens, you would do everything possible to persuade yourself that it didn't happen. From doubting the data, to doubting your perception and analysis.

Super-focus—Consider a tester whose job it is to test a specific requirement. While observing the system for that one requirement, other events or things may manifest. If the tester is super-focused on the task, he/she could easily miss even major extraneous observations. An interesting experiment called the monkey business illustrates this. You can see a video by Daniel J. Simons on YouTube at this web address: https://youtu.be/IGQmdoK_ZfY. There is also a related book with the title *The Invisible Gorilla* [37].

Confirmation bias—If we believe something to be true or perhaps want it to be true, we tend to seek/welcome information that confirms our belief, and dismiss the information that refutes our belief. This is a major factor in human existence. For example, you have heard the saying "love is blind," which describes a person who is in love and can see no faults or flaws in the beloved. This is a manifestation of

Safety Risk Management for Medical Devices
DOI: https://doi.org/10.1016/B978-0-12-813098-8.00034-9

© 2018 Elsevier Ltd.
All rights reserved.

confirmation bias, where only the information that supports the goodness of the beloved is accepted, and the information to the contrary is rejected.

Confirmation bias appears in science and engineering as well. The paper *False-Positive Psychology* by Simmons et al. [38] talks about biased selection and processing of test results to support hypotheses. Simmons says, "flexibility in data collection, analysis, and reporting dramatically increases actual false-positive rates." This is what is also referred to as "cherry picking." Some researchers have, in the past, selectively presented only the data that supported their claims and discarded the data that refuted their claims. This is what gave rise to the replication crisis in the early 2010s, as many scientific studies were difficult or impossible to reproduce in subsequent investigations.

Confirmation bias is also a reason why when an outlier happens in test results, attempts are made to find the root cause of the aberrant measurement, and discard it. But when the test results meet the expectations, no root-cause analysis is done.

Anchoring bias—You may have experienced that sometimes if you are not previously very confident in your thought or opinion, hearing someone else's thoughts sways you to their side. This is called anchoring bias. It is when another thought or piece of information anchors your thoughts and biases you toward the anchor. Imagine if you were going to guess the number of jelly beans in a jar to be 300. But before you could say anything, a respected, smarter person says there are at least 1000 jelly beans in that jar. Could you see yourself changing your estimate to a higher number?

In working meetings, typically a few people tend to dominate and anchor the thoughts of the other team members. Usually people with more authority or seniority have that power. Also, people with the loudest voices or those who are more self-assured or impassioned can anchor other people's thoughts.

Availability bias—When a thought or concept is more recent or easier to remember, it is seen as more true or more relevant. This is called the availability bias. For example, when a serious adverse event happens in the field, e.g., if a patient is seriously injured by a medical device, the whole engineering team and the management see that event as the highest risk that must be addressed. At the same time, it is possible that an even worse risk which has not happened yet, is lurking in the background.

The above examples of cognitive traps are some of the many ways that our minds can be led to make poor decisions. With respect to risk management, we need to be vigilant so that we do not miss Hazards, make good estimates of the risks, and make the best design decisions that reduce the risks of the medical devices. In your daily work, be mindful of cognitive biases that we all have, and try to be objective in considering and evaluating your own thoughts as well of the thoughts of others.

CHAPTER 35

Advice and Wisdom

Abstract

Mastery of the engineering and mathematics of risk management is not sufficient for success. Certain additional knowledge and experience helps propel the practitioner to success. In the closing chapter, advice and wisdom from 25^{+} years of experience in doing risk management in the medical device industry are presented as a complement to the knowledge that is presented in the rest of the chapters.

Keywords: Advice; wisdom

In closing, some advice and wisdom gathered over 25^{+} years of medical device industry experience:

- Team dynamics, how well the team members communicate, and whether there is a prevailing safety culture have an impact on the design and safety of the product.
- Discovery of a design flaw that could lead to a Hazard, could create an emotional reaction in the design engineers as a personal offense. Be sensitive and mindful of this possibility and couch your discoveries as opportunities for design improvement toward the shared goal of preservation of patient/user safety.
- Team continuity throughout the product lifecycle has an impact on product safety. It is possible that the original design team implemented certain safety features that were not well documented. Then during the maintenance part of the lifecycle, the continuation-engineering team may remove a safety feature due to lack of understanding of the rationale for the existence of that feature.
- Make the risk management file available to all team members. Access to this information helps the product development team make better decisions during the design and development phase.
- Write well! An easy to read and understand Risk Management Report will go a long way to build confidence and trust in a regulatory reviewer. Your colleagues will also appreciate well-written documents that are easy to understand and review.
- While compliance with ISO 14971 [3,7] offers a reasonable assurance of safety of the device, it does not mean the device will not cause harm.

Safety Risk Management for Medical Devices
DOI: https://doi.org/10.1016/B978-0-12-813098-8.00035-0

© 2018 Elsevier Ltd.
All rights reserved.

- Risk management is a living and an ongoing process for as long as the medical device is in the field.
- Include in the Risk Management Plan that a person with relevant medical knowledge will review and approve documents that evaluate Harms or risks of Harms, e.g., Harms Assessment List, Risk Assessment and Control Table, and benefit–risk analysis.
- Good traceability is an invaluable tool for both change-impact analysis, and determination of whether a detected field issue has a safety impact or not.
- When doing Use-Misuse Failure Modes and Effects Analysis, include in the team people with clinical knowledge, such as doctors, nurses, or field clinical engineers who spend time in the clinics/operating rooms and know how the product actually gets used.
- Near misses or close calls should not be dismissed. They are gifts—a warning without any harm. Learn from them as if they were real events.
- It is important to maintain consistency among the risk management artifacts and the design. As designs iterate, a robust change-control and configuration-management system helps ensure that the risk management documents remain consistent with the actual design of the medical device.
- Risk management produces large documents that are not easy to review. When asking people to review risk management documents, make specific requests of them to limit the scope of their work, e.g., ask the medical/clinical person to review the document from the medical safety perspective, while the mechanical engineer would evaluate the mechanical aspects in the document.
- Risk management is collaborative endeavor. In addition to the normal functions of R&D, manufacturing, Clinical, etc., engaging the Intended User may provide a unique perspective and may identify risks that relate to their expected use of the product in their specific environments.
- Even though the BXM method deploys a numerical method which lends itself to mathematical computations, one should not be lulled into thinking that the output of the analysis is the absolute truth. Remember that the input to the math is still an estimation. But it is better to make many small estimates and mathematically aggregate them, than to just make one big overall estimate. This is similar to estimating the annual budget of a company. The CFO doesn't just pull a number out of thin air. He/she asks people at all levels to make estimates for their budgets. Then these small estimates are gradually aggregated until the entire company budget is determined.
- It's important to cultivate and encourage humility in self and others. The absence of humility, can lead to unjustified certitude or hubris. This is when the spirit of inquiry stops and can lead to errors in risk management.
- Cultivate and encourage imagination.

APPENDIX A: GLOSSARY

ADE Adverse Device Effect
AE Adverse Event
AFAP As Far As Possible; equivalent to ALAP
ALAP As Low As Possible; equivalent to AFAP
ALARP As Low As Reasonably Practicable
Benefit In the context of risk management, benefit refers to Clinical Benefit and is defined as: positive impact or desirable outcome of a diagnostic procedure or therapeutic intervention on the health of an individual or a positive impact on patient management or public health. Benefits can be described in terms of magnitude, probability, and duration, among others [3]
BRA Benefit–Risk Analysis
CAPA Corrective and Preventive Actions
CCF Common Cause Failure
CFR Code of Federal Regulation
CHL Clinical Hazards List
CIP Clinical Investigation Plan
CRC Cyclic Redundancy Check
DFMEA Design Failure Modes and Effects Analysis
DMR Device Master Record
EMBASE Database published by Elsevier, contains over 11 million records with over 500,000 citations added annually. EMBASE's international journal collection contains over 5000 biomedical journals from 70 countries
Essential Design Output (EDO) Those design outputs that are essential to the proper functioning of the device – those functions that are related to safety and efficacy [21 CFR §820.30]
Essential Performance Performance of a clinical function, other than that related to basic safety, where loss or degradation beyond the limits specified by the manufacturer results in an unacceptable risk
Essential Requirements MDD/AIMDD Annex I. Requirements having to do with safe and effective performance of medical devices
EUDAMED The European information exchange system on Medical Devices
Failure Inability of an entity to achieve its purpose. This could be with no faults
Failure Mode The manner in which a product (system, subsystem, or component) can fail to perform its desired function, or meet its process or design requirements
Fault An anomalous condition for a part. Could result in failures
FMEA Failure Modes and Effects Analysis
FT Fault Tree
FTA Fault Tree Analysis
GUI Graphical User Interface
HAL Harms Assessment List
HHA, HHE Health Hazard (Analysis, Appraisal, Evaluation)
HiPPO Highest Paid Person in the Office
HS Hazardous Situation
IB Investigator's Brochure
IFU Information for Use
ISAO Information Sharing Analysis Organization
KOL Key Opinion Leader

Legacy Device Medical device which was legally placed on the market and is still marketed today but for which there is insufficient objective evidence that it is in compliance with the current version of ISO 14971

Legacy software MEDICAL DEVICE SOFTWARE which was legally placed on the market and is still marketed today but for which there is insufficient objective evidence that it was developed in compliance with the current version of this standard [9]

MDD Medical Device Directive

MDR Medical Device Reporting—Mandatory reporting requirement to the FDA

ME Medical Electrical equipment/systems

MedTech Medical Technology

NBRG Notified Bodies Recommendation Group

OR Operating Room

PFD Process Flow Diagram

PFMEA Process Failure Modes and Effects Analysis

PHA Preliminary Hazard Analysis

QC Quality Control

QMS Quality Management System

R&D Research and Development team/organization

RACT Risk Assessment and Control Table

RC Risk Control

RM Risk Management

RMF Risk Management File

RMT Risk Management Team

RPN Risk Priority Number

SADE Serious Adverse Device Effect

SAE Serious Adverse Event

SFMEA Software Failure Modes and Effects Analysis

SME Subject Matter Expert

Software defect An error in design/implementation of the software

Software failure A software condition that causes the System not to perform according to its specification

Software fault A software condition that causes the software not to perform as intended

Software Item Any identifiable part of a computer program, i.e., source code, object code, control code, control data, or a collection of these items [9]

Software System Integrated collection of Software Items organized to accomplish a specific function or set of functions [9]

Software Unit Software Item that is not subdivided into other items [9]

SOP Standard Operating Procedure

SOTA State of the Art

SOUP Software of Unknown Provenance

Standard of Care The level at which the average, prudent clinical care provider would practice

SW Software

TBD To Be Determined

UI User Interface

UMFMEA Use-Misuse Failure Modes and Effects Analysis

USA United States of America

Use Error User action or lack of user action while using the medical device that leads to a different result than that intended by the manufacturer or expected by the user [14]

VF Ventricular Fibrillation

APPENDIX B: TEMPLATES

In the following pages certain templates are provided as an aid to risk management practitioners.

Design Failure Modes and Effects Analysis (DFMEA) template
Software Failure Modes and Effects Analysis (SFMEA) template
Process Failure Modes and Effects Analysis (PFMEA) template
Use-Misuse Failure Modes and Effects Analysis (UMFMEA) template
Risk Assessment and Control Table (RACT) template

B.1 DESIGN FAILURE MODES AND EFFECTS ANALYSIS TEMPLATE

BXM

DFMEA
<insert subject of analysis>

Doc # 12345
Revision 1.0

Scope

This DFMEA covers the <insert subject of analysis> design.

The scope of the analysis is bounded in the diagram below and encompases all the items within the analysis boundary.

Item Under Analysis: <insert the subject of analysis>, version #.#

Primary functions: xxxx

Secondary functions: xxxx

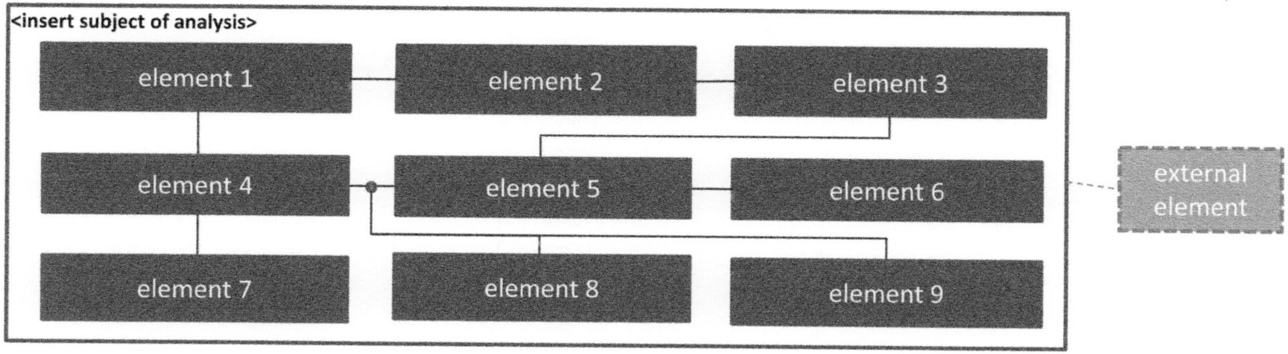

DFMEA Template - Copyright 2018 Bijan Elahi

BXM

DFMEA
<insert subject of analysis>

ITEM / FUNCTION				POTENTIAL FAILURE MODES & EFFECTS					Existing Mitigations	INITIAL RATING				Additional Mitigations	FINAL RATING				Remarks
ID	Source	Item	Function	Failure Mode	Causes /Mechanisms of Failure	Local Effects of Failure	End Effects of Failure	Safety Impact?		Sev	Occ	Det	RPN (auto)		Sev	Occ	Det	RPN (auto)	
1																			
2																			
3																			
4																			
5																			
6																			
7																			
8																			
9																			
10																			

DFMEA Template - Copyright 2018 Bijan Elahi

Severity Criteria (Sev)		
Rank	**Severity Descriptions - No Safety Impact**	**Severity Description – Safety Impact**
5	**Catastrophic**: Described failure mode will cause immediate failure of the Subject. (Total loss of all functions – primary and secondary)	**Catastrophic** – Impact of the end-effect at the System level can be death
4	**Critical**: Described failure mode will severely impact Subject functionality | Complete loss of primary functions	**Critical** – Impact of the end-effect at the System level can be permanent impairment or life-threatening injury
3	**Serious:** Described failure mode will reduce Subject functionality. (Partial loss of primary functions | Complete loss of secondary functions)	**Serious** – Impact of the end-effect at the System level can be injury or impairment that requires professional medical intervention
2	**Minor**: Described failure mode will have temporal or self-restoring impact on functionality | partial loss of secondary functions	**Minor** – Impact of the end-effect at the System level can be temporary injury or impairment that does not require professional medical intervention
1	**None**: Described component failure will have no impact on functionality	**Negligible** – Impact of the end-effect at the System level can be at most an inconvenience, or temporary discomfort

RPN	Action
53-125	**Level 3** - Reduce RPN through failure compensating provisions.
13-52	**Level 2** - If Safety Impact is Y, reduce RPN to as low as possible. If Safety Impact is N, reduce RPN if feasible.
1-12	**Level 1** - If Safety Impact is Y, reduce RPN to as low as possible. If Safety Impact is N, further RPN reduction is not required.

Probability of Occurrence Criteria (Occ)			
Category	**Rank**	**Qualitative Criteria**	**Quantitative Criteria**
Frequent	5	The occurrence is frequent. Failure may be almost certain or constant failure.	$\geq 10^{-3}$
Probable	4	The occurrence is probable. Failure may be likely | repeated failures are expected.	$< 10^{-3}$ and $\geq 10^{-4}$
Occasional	3	The occurrence is occasional. Failures may occur at infrequent intervals.	$< 10^{-4}$ and $\geq 10^{-5}$
Remote	2	The occurrence is remote. Failures are seldom expected to occur.	$< 10^{-5}$ and $\geq 10^{-6}$
Improbable	1	The occurrence is improbable. The failure is not expected to occur.	$< 10^{-6}$

Detection Criteria (Det)			
Category	**Rank**	**Qualitative Criteria**	**Quantitative Criteria**
Undetectable	5	No detection opportunity | No means for detection | Countermeasures not possible	$< 10^{-3}$
Low	4	Opportunity for detection is low | Countermeasures are unlikely	$< 10^{-2}$ and $\geq 10^{-3}$
Moderate	3	Opportunity for detection is moderate | Countermeasures are probable	$< 10^{-1}$ and $\geq 10^{-2}$
High	2	Opportunity for detection is high | Countermeasures are likely	$< 9 \times 10^{-1}$ and $\geq 10^{-1}$
Almost Certain	1	Opportunity for detection is almost certain | Countermeasures are certain	$\geq 9 \times 10^{-1}$

DFMEA Template - Copyright 2018 Bijan Elahi

BXM

DFMEA
<insert subject of analysis>

Doc # 12345
Revision 1.0

Revision History

Revision	Author	CR	Description of Change

DFMEA Template - Copyright 2018 Bijan Elahi

DFMEA - <insert subject of analysis>
Log of Working Sessions

Doc # 12345
Revision 1.0

Date	Participants

DFMEA Template- Copyright 2018 Bijan Elahi

B.2 SOFTWARE FAILURE MODES AND EFFECTS ANALYSIS TEMPLATE

BXM

SFMEA

<insert subject of analysis>

Doc # 12345

Revision 1.0

Scope

This SFMEA covers <insert subject of analysis> design.

The scope of the analysis is bounded in the diagram below and encompases all the items within the analysis boundary.

Item Under Analysis: <insert the subject of analysis>, version #.#

Primary functions: xxxx

Secondary functions: xxxx

<replace the example graphic below with a diagram suitable for your analysis>

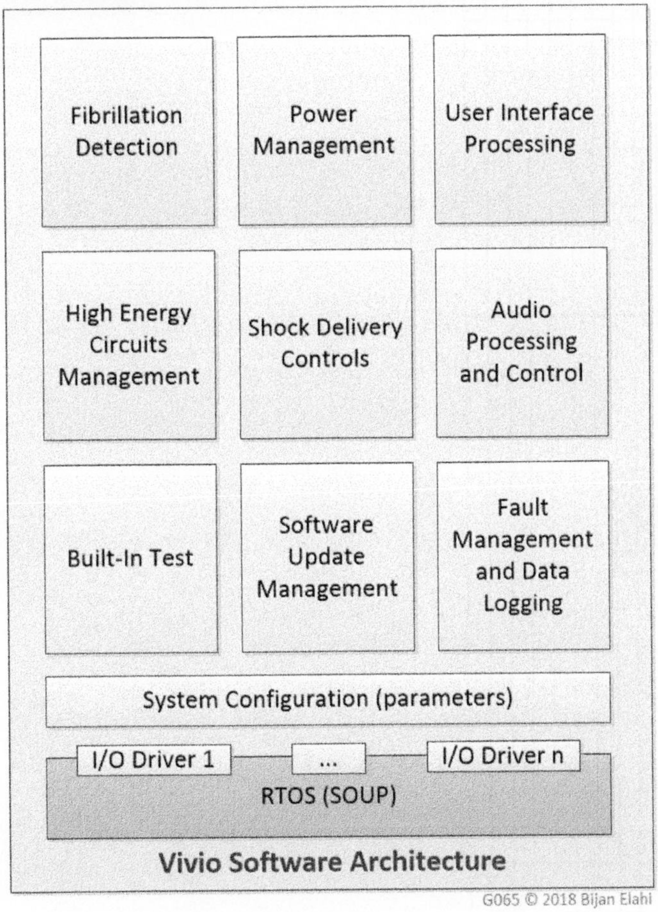

G065 © 2018 Bijan Elahi

BXM

SFMEA
<insert subject of analysis>

Doc # 12345
Revision 1.0

ITEM / FUNCTION				POTENTIAL FAILURE MODES & EFFECTS						INITIAL RATING						FINAL RATING					
ID	Source	Item	Function	Failure Mode	Causes /Mechanisms of Failure	Local Effects of Failure	End Effects of Failure	Safety Impact?	Existing Mitigations	Sev	Occ	Det	RPN (auto)	Crit (auto)	Additional Mitigations	Sev	Occ	Det	RPN (auto)	Crit (auto)	Remarks
1																					
2																					
3																					
4																					
5																					
6																					
7																					
8																					
9																					
10																					

SFMEA Template - © 2018 Bijan Elahi

Severity Criteria (Sev)		
Rank	**Severity Descriptions - No Safety Impact**	**Severity Description -- Safety Impact**
5	**Catastrophic:** Described failure mode will cause immediate failure of the Subject. (Total loss of all functions – primary and secondary)	**Catastrophic** – Impact of the end-effect at the System level can be death
4	**Critical:** Described failure mode will severely impact Subject functionality | Complete loss of primary functions	**Critical** – Impact of the end-effect at the System level can be permanent impairment or life-threatening injury
3	**Serious:** Described failure mode will reduce Subject functionality. (Partial loss of primary functions | Complete loss of secondary	**Serious** – Impact of the end-effect at the System level can be injury or impairment that requires professional medical intervention
2	**Minor:** Described failure mode will have temporal or self-restoring impact on functionality | partial loss of secondary functions	**Minor** – Impact of the end-effect at the System level can be temporary injury or impairment that does not require professional medical
1	**None:** Described component failure will have no impact on functionality	**Negligible** – Impact of the end-effect at the System level can be at most an inconvenience, or temporary discomfort

RPN	Action
53-125	**Level 3** - Reduce RPN through failure compensating provisions.
13-52	**Level 2** – If Safety Impact is Y, reduce RPN to as low as possible. If Safety Impact is N, reduce RPN if feasible.
1-12	**Level 1** - If Safety Impact is Y, reduce RPN to as low as possible. If Safety Impact is N, further RPN reduction is not required.

Criticality		Severity				
		1	2	3	4	5
Detectability	5	2	2	3	3	3
	4	1	2	2	3	3
	3	1	1	2	2	3
	2	1	1	1	2	3
	1	1	1	1	1	2

Probability of Occurrence Criteria (Occ)			
Category	**Rank**	**Qualitative Criteria**	**Quantitative Criteria**
Frequent	5	The occurrence is frequent. Failure may be almost certain or constant failure.	$\geq 10^{-3}$
Probable	4	The occurrence is probable. Failure may be likely | repeated failures are expected.	$< 10^{-3}$ and $\geq 10^{-4}$
Occasional	3	The occurrence is occasional. Failures may occur at infrequent intervals.	$< 10^{-4}$ and $\geq 10^{-5}$
Remote	2	The occurrence is remote. Failures are seldom expected to occur.	$< 10^{-5}$ and $\geq 10^{-6}$
Improbable	1	The occurrence is improbable. The failure is not expected to occur.	$< 10^{-6}$

Detection Criteria (Det)			
Category	**Rank**	**Qualitative Criteria**	**Quantitative Criteria**
Undetectable	5	No detection opportunity | No means for detection | Countermeasures not possible	$< 10^{-3}$
Low	4	Opportunity for detection is low | Countermeasures are unlikely	$< 10^{-2}$ and $\geq 10^{-3}$
Moderate	3	Opportunity for detection is moderate | Countermeasures are probable	$< 10^{-1}$ and $\geq 10^{-2}$
High	2	Opportunity for detection is high | Countermeasures are likely	$< 9 \times 10^{-1}$ and $\geq 10^{-1}$
Almost Certain	1	Opportunity for detection is almost certain | Countermeasures are certain	$\geq 9 \times 10^{-1}$

BXM

SFMEA
<insert subject of analysis>

Doc # 12345
Revision 1.0

Revision History

Revision	Author	CR	Description of Change

SFMEA Template - © 2018 Bijan Elahi

BXM

SFMEA - <insert subject of analysis>
Log of Working Sessions

Doc # 12345
Revision 1.0

Date	Participants

B.3 PROCESS FAILURE MODES AND EFFECTS ANALYSIS TEMPLATE

BXM

PFMEA
<insert process name>

Doc # 12345
Revision 1.0

Scope

This PFMEA covers the manufacturing process for <insert product of the process>.

The scope of the analysis is bounded in the diagram below and encompases all the items within the analysis boundary.

Process Under Analysis: Manufacturing process xxxx, for <insert product of the process>, version #.#

Primary functions: xxxx

Secondary functions: xxxx

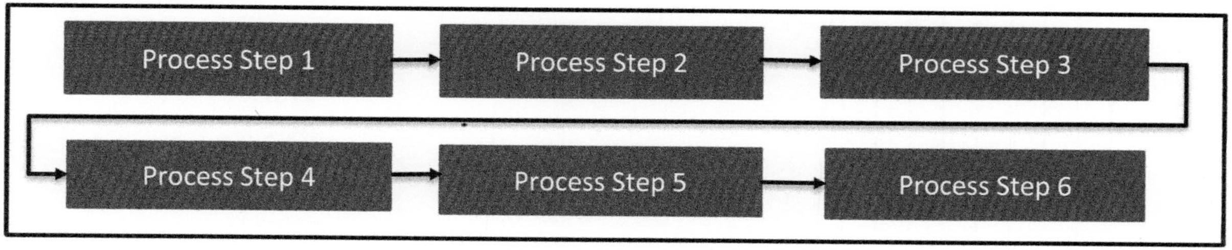

<add drawings, or pictures to help the reader understand the process under analysis>

PFMEA Template - Copyright Bijan Elahi 2018

ITEM / FUNCTION			POTENTIAL FAILURE MODES & EFFECTS						INITIAL RATING					FINAL RATING				
ID	Process Step	Process Step Function	Failure Mode	Causes/Mechanisms of Failure Mode	Local Effects of Failure Mode	End Effects of Failure Mode	Safety Impact?	Existing Mitigations	Sev	Occ	Det	RPN (auto)	Additional Mitigations	Sev	Occ	Det	RPN (auto)	Remarks
Task 1 - <insert task name/description>																		
1																		
2																		
3																		
4																		
5																		
6																		
7																		
8																		
9																		
10																		
11																		

PFMEA Template - Copyright 2018 Bijan Elahi

Severity Criteria (Sev)		
Rank	Severity Descriptions - No Safety Impact	Severity Description -- Safety Impact
5	Failure to meet Regulatory requirements | Total loss of all functions – primary and secondary | >70% of the production has to be scrapped	**Catastrophic** – Impact of the end-effect at the System level can be death
4	Loss or degradation of primary functions | Failure to meet product specification | Scrapping of 50-70% of the production	**Critical** – Impact of the end-effect at the System level can be permanent impairment or life-threatening injury
3	Loss or degradation of secondary functions | Reduced reliability but still within Spec | Scrapping of 25-50% of the production.	**Serious** – Impact of the end-effect at the System level can be injury or impairment that requires professional medical
2	Process Delay | Scrapping of 5-25% of the production. | Minor cosmetic or usability impact but still within Spec	**Minor** – Impact of the end-effect at the System level can be temporary injury or impairment that does not require professional medical intervention
1	Scrapping of 0-5% of the production. Some of the products have to be reworked	**Negligible** – Impact of the end-effect at the System level can be at most an inconvenience, or temporary discomfort

RPN	Action
53-125	**Level 3** - Reduce RPN through failure compensating provisions.
13-52	**Level 2** - If Safety Impact is Y, reduce RPN to as low as possible. If Safety Impact is N, reduce RPN if feasible.
1-12	**Level 1** - If Safety Impact is Y, reduce RPN to as low as possible. If Safety Impact is N, further RPN reduction is not required.

Probability of Occurrence Criteria (Occ)			
Category	Rank	Qualitative Criteria	Quantitative Criteria
Frequent	5	The occurrence is frequent. Failure may be almost certain or constant failure.	$\geq 10^{-1}$
Probable	4	The occurrence is probable. Failure may be likely | Repeated failures are expected.	$< 10^{-1}$ and $\geq 10^{-2}$
Occasional	3	The occurrence is occasional | Failures may occur at infrequent intervals.	$< 10^{-2}$ and $\geq 10^{-3}$
Remote	2	The occurrence is remote | Failures are seldom expected to occur.	$< 10^{-3}$ and $\geq 10^{-4}$
Improbable	1	The occurrence is improbable | Failure is not expected to occur.	$< 10^{-4}$

Detection Criteria (Det)			
Category	Rank	Qualitative Criteria	Quantitative Criteria
Undetectable	5	Physics-of-Failure not understood | No detection opportunity (e.g. no inspection, or no means for detection) | Countermeasures not possible	$< 10^{-3}$
Low	4	Failure is very difficult to detect | Opportunity for detection is low (e.g., very low sampling for inspection) | Countermeasures are unlikely	$< 10^{-2}$ and $\geq 10^{-3}$
Moderate	3	Failure is moderately detectable | Opportunity for detection is moderate (e.g. 10% sampling) | Countermeasures are probable	$< 10^{-1}$ and $\geq 10^{-2}$
High	2	Failure is very detectable | Opportunity for detection is high (e.g. 100% visual inspection) | Countermeasures are likely	$< 9 \times 10^{-1}$ and $\geq 10^{-1}$
Almost Certain	1	Failure is obvious | Opportunity for detection is almost certain (e.g., 100% instrumented inspection) | Countermeasures are certain	$\geq 9 \times 10^{-1}$

BXM

PFMEA

<insert process name>

Doc # 12345

Revision 1.0

Revision History

Revision	Author	CR	Description of Change

BXM

PFMEA - <enter process name>
Log of Working Sessions

Doc # 12345
Revision 1.0

Date	Participants

PFMEA Template - Copyright Bijan Elahi 2018

B.4 USE-MISUSE FAILURE MODES AND EFFECTS ANALYSIS TEMPLATE

BXM

UMFMEA
<insert process name>

Doc # 12345
Revision 1.0

Introduction

Use-Misuse Failure Modes and Effects Analysis (UMFMEA) analyses failures that are related to use by the User. UMFMEA also considers potential misuses. Abnormal use or malice are excluded.

Reasonably Foreseeable Misuses are also analyzed in this analysis. Misuse is not use failure. It is deliberate and well-intentioned. Example: Off-label use

System Under Analysis:

<enter the name and version number of the system under analysis>

Primary functions:

xxxx

Secondary functions:

xxxx

UMFMEA Template - Copyright 2018 Bijan Elahi

UMFMEA
<insert process name>

Doc # 12345
Revision 1.0

Scope

<Describe the scope of analysis: the system and all the actors>

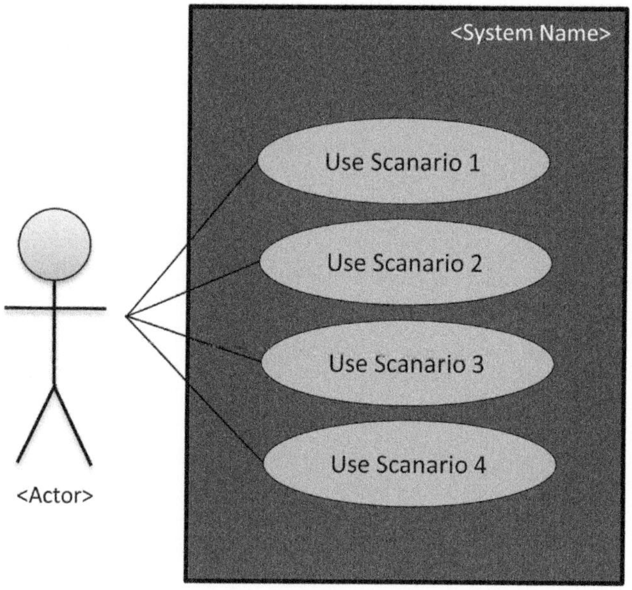

UMFMEA Template - Copyright 2018 Bijan Elahi

BXM

Use Scenario		POTENTIAL FAILURE MODES & EFFECTS						INITIAL RATING					FINAL RATING				
ID	Step Action	Failure Mode	Causes /Mechanisms of Failure	Local Effects of Failure	End Effects of Failure	Safety Impact?	Existing Mitigations	Sev	Occ	Det	RPN (auto)	Additional Mitigations	Sev	Occ	Det	RPN (auto)	Remarks
Use Scenario 1 - xxxx																	
Task 1 - xxxx																	
1																	
2																	
Task 2 - xxxx																	
3																	
4																	
5																	
Misuses																	
6																	

UMFMEA Template - Copyright 2018 Bijan Elahi

<table>
<tr><th colspan="3">Severity Criteria (Sev)</th></tr>
<tr><th>Rank</th><th>Severity Descriptions - No Safety Impact</th><th>Severity Description -- Safety Impact</th></tr>
<tr><td>5</td><td>Catastrophic: Described failure mode will cause immediate failure of the Subject. (Total loss of all function – primary and secondary)</td><td>Catastrophic: Impact of the end-effect at the System level can be death</td></tr>
<tr><td>4</td><td>Critical: Described failure mode will severely impact Subject functionality | Complete loss of primary functions</td><td>Critical: Impact of the end-effect at the System level can be permanent impairment or life-threatening injury</td></tr>
<tr><td>3</td><td>Serious: Described failure mode will reduce Subject functionality. (Partial loss of primary functions | Complete loss of secondary functions)</td><td>Serious: Impact of the end-effect at the System level can be injury or impairment that requires professional medical intervention</td></tr>
<tr><td>2</td><td>Minor: Described failure mode will have temporal or self-restoring impact on functionality | partial loss of secondary functions</td><td>Minor: Impact of the end-effect at the System level can be temporary injury or impairment that does not require professional medical intervention</td></tr>
<tr><td>1</td><td>None: Described failure mode will have no impact on functionality | Annoyance / inconvenience of the user</td><td>Negligible: Impact of the end-effect at the System level can be at most an inconvenience, or temporary discomfort</td></tr>
</table>

Probability of Occurrence Criteria (Occ)		
Category	**Rank**	**Qualitative Criteria**
Frequent	5	The occurrence is frequent. Experienced by almost every user.
Probable	4	The occurrence is probable. Experienced by most users.
Occasional	3	The occurrence is occasional. Experienced by some users.
Remote	2	The occurrence is remote. Experienced by few users.
Improbable	1	The occurrence is improbable. Has not been observed; not expected to be experienced by any user.

RPN	Action
53-125	**Level 3** - Reduce RPN through failure compensating provisions.
13-52	**Level 2** - If Safety Impact is Y, reduce RPN to as low as possible. If Safety Impact is N, reduce RPN if feasible.
1-12	**Level 1** - If Safety Impact is Y, reduce RPN to as low as possible. If Safety Impact is N, further RPN reduction is not required.

<table>
<tr><th colspan="3">Detection Criteria (Det)</th></tr>
<tr><th>Category</th><th>Rank</th><th>Descriptions</th></tr>
<tr><td>Undetectable</td><td>5</td><td>Effect is not immediately visible or knowable | Countermeasures not possible</td></tr>
<tr><td>Low</td><td>4</td><td>Effect can be visible or knowable only with expert investigation using specialized equipment | Countermeasures are unlikely</td></tr>
<tr><td>Moderate</td><td>3</td><td>Effect can be visible or knowable with the moderate effort by user | Countermeasures are probable</td></tr>
<tr><td>High</td><td>2</td><td>Highly Detectable - Effect can be visible or knowable with simple action by user, from the information provided by the system itself | Countermeasures are likely</td></tr>
<tr><td>Almost Certain</td><td>1</td><td>Almost certain detection - Effect is clearly visible or knowable to user without any further action by user | Countermeasures are certain</td></tr>
</table>

UMFMEA Template - Copyright 2018 Bijan Elahi

BXM

UMFMEA
<enter product name>

Doc # 12345
Revision 1.0

Revision History

Revision	Author	CR	Description of Change

UMFMEA Template - Copyright 2018 Bijan Elahi

BXM

UMFMEA - <enter product name>
Log of Working Sessions

Doc # 12345
Revision 1.0

Date	Participants

UMFMEA Template - Copyright 2018 Bijan Elahi

B.5 RISK ASSESSMENT AND CONTROL TABLE TEMPLATE

BXM

RACT
<insert product name>

Doc # 12345
Revision 1.0

ID	Hazard Source	Initial cause of hazard	Sequence of Events	Hazard	Hazardous Situations	Risk Controls			P1	Harm	P2					Risk				
						SD	PM	IS			Cat	Crit	Ser	Minr	Negl	Cat	Crit	Ser	Minr	Negl
1																0.00E+00	0.00E+00	0.00E+00	0.00E+00	0.00E+00
2																0.00E+00	0.00E+00	0.00E+00	0.00E+00	0.00E+00
3																0.00E+00	0.00E+00	0.00E+00	0.00E+00	0.00E+00
4																0.00E+00	0.00E+00	0.00E+00	0.00E+00	0.00E+00
5																0.00E+00	0.00E+00	0.00E+00	0.00E+00	0.00E+00
6																0.00E+00	0.00E+00	0.00E+00	0.00E+00	0.00E+00
7																0.00E+00	0.00E+00	0.00E+00	0.00E+00	0.00E+00
8																0.00E+00	0.00E+00	0.00E+00	0.00E+00	0.00E+00
9																0.00E+00	0.00E+00	0.00E+00	0.00E+00	0.00E+00
10																0.00E+00	0.00E+00	0.00E+00	0.00E+00	0.00E+00

BXM

RACT - <insert product name>
Acceptable Risk Limits

Doc # 12345
Revision 1.0

<enter the source of acceptable risk limits, e.g., published scientific papers>

R-Cat	R-Crit	R-Ser	R-Minr	R-Negl
0.0E+00	0.0E+00	0.0E+00	0.0E+00	0.0E+00

RACT Template - Copyright 2017 Bijan Elahi

BXM

RACT
<insert product name>

Doc # 12345
Revision 1.0

Abbreviations

Term	Definition
Cat	Catastrophic
Crit	Critical
ID	Identification
IS	Information for Safety
Minr	Minor
Negl	Negligible
PM	Protective Measure
RACT	Risk Assessment and Control Table
SD	Safe by Design
Ser	Serious

Revision History

Revision	Author	CR	Description of Change

BXM

<insert product name>
Log of Working Sessions

Doc # 12345
Revision 1.0

Date	Participants

RACT Template - Copyright 2017 Bijan Elahi

Page 4 of 4

APPENDIX C: EXAMPLE DEVICE—VIVIO

In this appendix, the BXM method is applied to a hypothetical Automatic External Defibrillator (AED) named Vivio. The purpose for this example is to teach by illustration, the mechanics of the BXM method. Vivio is not a real device. The cited electronics and mechanical designs, or the Failure Modes or mitigations are not from a real device. Use the example as a vehicle to learn how risk management is done per the BXM method, and how the different elements of risk management are connected to one another.

The Vivio example is deliberately simplified and abbreviated to ease comprehension by the reader, and also to fit within the bounds of this book.

- Only *System-level* Failure Modes and Effects Analyses (FMEAs) are provided: System Design Failure Modes and Effects Analysis (DFMEA), System Process Failure Modes and Effects Analysis (PFMEA), and Use-Misuse Failure Modes and Effects Analysis (UMFMEA). Lower level FMEAs are presumed to be performed as needed.
- Only some of the Failure Modes with safety impact are carried forward to the Risk Assessment and Control Table (RACT). In a real project all Failure Modes in System-level FMEAs that have a safety impact should be carried forward to the RACT.

C.1 VIVIO PRODUCT DESCRIPTION

Vivio is an AED. It is small, lightweight, rugged, and battery operated. Vivio is intended for simple and reliable operation by minimally trained people.

Vivio is intended to treat Ventricular Fibrillation (VF), the most common cause of sudden cardiac death (SCD). VF is a chaotic quivering of the heart muscle that prevents the heart from pumping blood. The only effective treatment of VF is defibrillaiton, which is sending an electric shock across the heart, so as to reset the heart and enable it to start pumping again. A victim of VF is genrally unconscious. When properly applied, Vivio automatically senses, and diagnoses the victim's heart rhythm, and if VF is detected, Vivio delivers a high voltage electric shock to the victim's heart.

Vivio is comprised of the base unit, two disposable pads, and the outer casing. The user is expected to apply the pads to the victim's bare chest according to the drawings on the pads, and follow the verbal instructions given by the base unit.

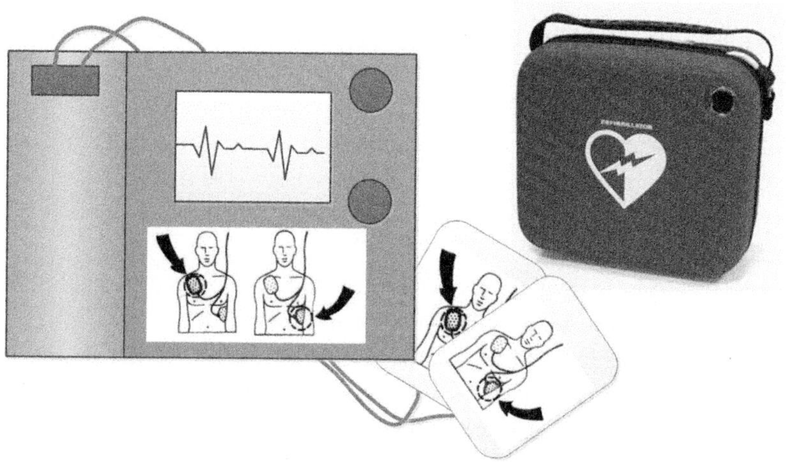

C.2 VIVIO PRODUCT REQUIREMENTS

The following are the System requirement specifications for Vivio:

- Automatically detect ventricular fibrillation with a sensitivity of $\geq 95\%$ and specificity of $\geq 90\%$
- Deliver at least 150 joules of energy per shock
- Able to deliver a minimum of 200 shocks on a fresh battery
- Standby longevity ≥ 4 years on a fresh battery
- Max time between shocks: 20 seconds
- Able to detect and cope with artifacts, e.g., implanted pacemakers
- Visible indication of readiness for use
- Protective case for installation in public spaces
- Weight: ≤ 2 kg (with battery, case, and pads)
- Resistance to dust and water ingress:
  - While inside the case: IP54 or better
  - While outside the case: IP52 or better

- Environmental constraints:
  - Temperature: 0–50°C
  - Altitude: 0–4500 m above sea level
  - Crush: 225 kg

C.3 VIVIO ARCHITECTURE

The high-level architectural design of Vivio is depicted in Fig. C.1. The main blocks of Vivio design are identified in order to aid in the understanding of the product function and also to facilitate performing FMEAs on Vivio.

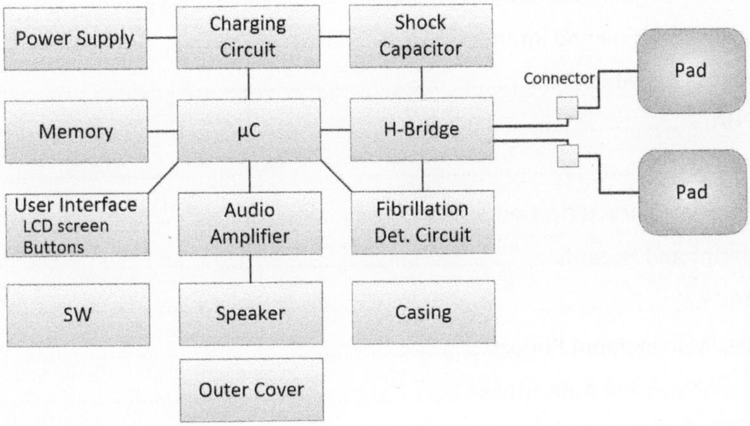

Figure C.1 Vivio system architecture.

C.4 RISK MANAGEMENT PLAN

BXM	**Risk Management Plan – Vivio AED**	Doc # 12340 Revision 1.0 Page 1 of 11

Table of Contents

Example RMP – Copyright 2018 Bijan Elahi

BXM	Risk Management Plan – Vivio AED	Doc # 12340 Revision 1.0 Page 2 of 11

1 INTRODUCTION

This document is the risk management plan for Vivio, Automatic External Defibrillator.

1.1 Purpose

This Risk Management Plan (RMP) describes the planned safety Risk Management Activities for the Vivio Project following [RM SOP]. The RMP addresses:

a) the scope of the planned Risk Management activities, identifying and describing the Vivio System and the life-cycle phases for which each element of the plan is applicable;
b) assignment of responsibilities and authorities;
c) requirements for review of risk management activities;
d) criteria for risk acceptability, including criteria for accepting risks when the probability of occurrence of harm cannot be estimated;
e) verification activities to:
   - e.1) verify the implementation of risk controls;
   - e.2) verify the effectiveness of risk controls, e.g. by collection of clinical data, or usability studies;
f) activities related to collection and review of relevant production and post-production information

1.2 Scope

The scope of this plan is from Concept-Release to End of Verification phases of the product lifecycle. This plan will be updated by the End of Verification phase to account for the remaining phases of the product development lifecycles.

Some of the activities in support of Risk Management, e.g. biocompatibility testing, or sterilization are out of scope of this plan. However, the safety impact of sterility and biocompatibility will be captured in the RACT.

Example RMP – Copyright 2018 Bijan Elahi

BXM	Risk Management Plan – Vivio AED	Doc # 12340 Revision 1.0 Page 3 of 11

The Vivio Automatic External Defibrillator System is comprised of several sourced components. The quality agreements with the suppliers of said components stipulate the required risk management inputs from the suppliers.

1.3 Definitions & Abbreviations

See [Glossary] for an overview of definitions & abbreviations used in this document.

The following definitions will be used in the RM process:

Table 1 - Definitions of Severity Classes (ISO 14971 Table D.3)

Class	Definition
Catastrophic	Patient or operator death
Critical	Permanent impairment or life-threatening injury
Serious	Injury or impairment that requires professional medical intervention
Minor	Temporary injury or impairment not requiring professional medical intervention
Negligible	Inconvenience, or temporary discomfort

Example RMP – Copyright 2018 Bijan Elahi

BXM	Risk Management Plan – Vivio AED	Doc # 12340 Revision 1.0 Page 4 of 11

2 PRODUCT DESCRIPTION AND INTENDED USE

Vivio is a portable Automatic External Defibrillator (AED), designed for outdoor storage, and use by minimally trained individuals. Vivio is equipped with a color LCD display. Vivio provides use instructions in graphic format on the LCD display, and speaks in 14 languages. Language selection is configurable for use in various geographies.

When contact-pads are properly applied to the patient chest, Vivio automatically monitors patient's heart rhythm, detects fibrillation and emits a bi-phasic trans-thoracic shock of approximately 360 J to defibrillate the heart. Vivio operates on a primary battery, which is user-replaceable. Contact-pads are single-use products which are supplied by a third-party manufacturer. Vivio is intended for use on adults, and children who are over 25 kg, or older than 8 years.

3 RM PROCESS ACTIVITIES

3.1 PHA

At the concept phase, a Preliminary Hazard Analysis (PHA) will be performed for a high-level assessment of the System risks. The PHA will include a Fault Tree Analysis.

3.2 FMEA

The project will execute Design, Use/Misuse, and Process FMEAs according to [RM SOP]. Software FMEA will be performed subordinate to System DFMEA. At the System level, a DFMEA, PFMEA and UMFMEA will be produced.

FMEA's are conducted with the intention to identify potential causal chains that could lead to System hazards. During FMEAs Failure Modes whose End Effects do not lead to System hazards may also be identified. End effects from top-level FMEAs that could

Example RMP – Copyright 2018 Bijan Elahi

potentially do lead to harms will be captured as hazards in the [RACT] for risk assessment.

3.3 Product Characterization

The risk management team will characterize Vivio using the questions listed in annex C of ISO 14971. The results of annex C answers will be logged in the risk management file.

3.4 Harms and Hazards

A standardized set of potential harms related to the use of Vivio have been identified in [HAL]. For each of these harms, the probabilities of occurrence of the outcome from that harm have been estimated in five severity classes as defined in **Table 1** above.

3.5 RACT

The results of lower-level FMEAs will be rolled up into a System DFMEA. The RACT will be populated from the hazards that are identified in the top-level System FMEAs, and risks for each hazard, hazardous situation and overall for Vivio will be computed and evaluated for acceptability.

3.6 Risk Management Report

A summary of the risk management activities and the conclusion of product risks and benefits will be captured in the Risk Management Report [RMR].

Example RMP – Copyright 2018 Bijan Elahi

BXM	Risk Management Plan – Vivio AED	Doc # 12340 Revision 1.0 Page 6 of 11

4 RESPONSIBILITIES AND AUTHORITIES

The composition of Vivio Risk Management Team is detailed in the table below:

Role	Responsibilities	Requirements for Review of RM Activities
Management	Provided resources and qualified personnel	Review of RMP and RMR.
Risk Manager	Ensure RM activities and work products are in compliance with [RM SOP] Maintain the Risk Management File	Review all RM work products
Quality	Ensure RM activities and work products are in compliance with "Company" QMS	Review all RM work products with respect to Quality responsibilities
Regulatory	Ensure RM work products meet applicable standards and directives. Ensure all work products are appropriate for regulatory authority submission.	Review all RM work products with respect to Regulatory responsibilities

BXM	Risk Management Plan – Vivio AED	Doc # 12340 Revision 1.0 Page 7 of 11

Role	Responsibilities	Requirements for Review of RM Activities
Systems Engineering	Ensure RM work products are consistent with the system design and risk control measures are valid, achievable and reasonable. Ensure RM work products meet applicable standards. Ensure the RM contributions by external suppliers are according to quality agreements.	Review all RM work products with respect to Systems responsibilities
Medical/Clinical	Ensure that the RM process makes sound choices from a medical perspective and that interactions with patient/user are correctly considered	Review all RM work products with respect to Medical/Clinical responsibilities
SMEs	Provide expert opinion/consulting to RMTas needed to assist in technical evaluation of RM work products Ensure RM work products meet applicable standards, as appropriate.	No formal review responsibilities

Prior to integration into the RACT, the input from suppliers will be reviewed and approved by Systems engineering as well as the RMT as defined in the table above.

Example RMP – Copyright 2018 Bijan Elahi

BXM	Risk Management Plan – Vivio AED	Doc # 12340 Revision 1.0 Page 8 of 11

4.1 Review of Risk Management Activities

Risk management activities will be subject to review as part of the normal flow of product development. At design process milestones, as defined in the [PDP SOP], Project Management ensures completion of reviews of the RM activities that are due by that milestone.

5 RISK ACCEPTANCE CRITERIA

Using the methodology that is described in [RACT Guidance] residual risks will be computed per hazard, per hazardous situation, and overall. Risks will be considered acceptable based on the following criteria in the order of priority:

Compliance with EU harmonized standards

Compliance with other national or international standards (non-harmonized, but accepted in the industry as applicable)

Comparison with historical data and best medical practices, i.e. state-of-the-art

Table 2 shows the criteria for overall residual risk evaluation. The overall residual risks of the Vivio System must be less than, or equal to the values cited in Table 2 for all severity classes, and be reduced to as low as possible.

Table 2 - Overall Residual Risk

Overall Residual Risk Evaluation	Severity Category				
	Negligible	Minor	Serious	Critical	Catastrophic
Reference Risk Levels	$\leq 1.0\text{x}10^{-2}$	$\leq 1.5\text{x}10^{-3}$	$\leq 1.3\text{x}10^{-4}$	$\leq 9.5\text{x}10^{-5}$	$\leq 7.1\text{x}10^{-5}$

Note - In this RM process, a clearly non-functional device is not considered to be a hazard. This includes: a device that has a dead battery and cannot be even turned on; or a device that is warning that it is non-functional; or a device that defective and has not left the factory yet. Rationale is that such a device would not be put to use and therefore would not expose the patient to hazard(s). However, a device that is expected to be functional, but doesn't perform as expected, is considered to be a hazard.

6 RISK BENEFIT ANALYSIS

A risk benefit analysis will be done for individual risks, and the overall residual risks.

7 RISK CONTROL STRATEGY

Risk control will be done per [RM SOP] par 6.4. Labelling in the form of informing the users of residual risk of Vivio will not be used as risk control. However, information for safety in the form of instructions for safe and proper use of the system may be used to control risks.

8 RISK MANAGEMENT FILE

Per [RM SOP], a Risk Management File (RMF) will be created and maintained. The RMF will be a part of the DHF.

The RMF Content will be:

- This plan, including the residual risk acceptance criteria
- Harms Assessment List [HAL]
- Clinical Hazard List [CHL]
- Preliminary Hazard Analysis [PHA]
- FMEA reports
- Risk Assessment and Control Table [RACT]
- Risk management report(s) [RMR]
- Risk Controls verification reports
- Records of reviews and approvals of RM artifacts

Example RMP – Copyright 2018 Bijan Elahi

BXM	Risk Management Plan – Vivio AED	Doc # 12340 Revision 1.0 Page 10 of 11

9 VERIFICATION OF RISK CONTROLS

Risk controls will be verified for implementation and effectiveness. The result of this activity will be documented and stored in the RMF. It may be possible in some cases, to combine the verification of implementation and effectiveness of some risk controls in the same test.

10 PRODUCTION AND POST-PRODUCTION ACTIVITIES

Information from production that is relevant to safety will be fed back to the risk management process on a quarterly basis. For Post-Production Information, per [RM SOP] the Complaint Handling, Vigilance, and Postmarket Surveillance processes will be used to collect field information about Vivio AED. The sources of post-production input can be: Manufacturing, R&D, Sales, Marketing, Customers, Patients, Distributors, postmarket clinical trials, published scientific papers, news media, adverse event reports – including for competitive products. On an annual basis, or more frequently if a significant discovery is made, data from above sources will be evaluated for relevance to Vivio AED. The ensuing actions depend on the collected information and can fall in a spectrum, including:

- Documentation of the information-collection actions, and discoveries where no change to the Vivio AED Risk Management artifacts is necessary
- Update to Vivio AED [RMF] including FMEAs, [HAR] and RMR with outcomes being:
    - Overall residual risk remains acceptable and benefits outweigh risks
    - Overall residual risk no longer acceptable, triggering a range of other actions e.g. Health Hazard Assessment, CAPAs, Field Corrective Actions, Vigilance reporting, redesign, etc.

Also, based on the new knowledge gained from post-production information [CHL], [HAL], or [RM SOP] may be updated.

Example RMP – Copyright 2018 Bijan Elahi

BXM	Risk Management Plan – Vivio AED	Doc # 12340 Revision 1.0 Page 11 of 11

11 REFERENCES

Reference	Document number	Title
[CHL]	12342	Clinical Hazards List
[Glossary]	xxxxxx	QMS Glossary for "Company"
[HAL]	12343	Harms Assessment List
[PDP SOP]	xxxxxx	Product Devlopment Process Standard Operating Procedure
[PHA]	12341	Preliminary Hazard Analysis
[RACT Guidance]	xxxxxx	Guidance document for the creation of RACT
[RACT]	12347	Risk Assessment and Control Table
[RM SOP]	xxxxxx	Risk Management Standard Operating Procedure
[RMR]	12349	Risk Management Report

12 REVISION HISTORY

Revision	Author	CR	Description of changes
1.0	John Adams	N/A	First approved version

Example RMP – Copyright 2018 Bijan Elahi

C.5 CLINICAL HAZARDS LIST

BXM	Clinical Hazards List	Doc # 12342 Revision 1.0 Page 1 of 2

1 INTRODUCTION

This document embodies the list of clinical hazards that are common to all the products that are designed, developed or produced by "Company". The sources of the information for this document are:

- Literature search of comparable products
- ISO 14971, especially the risk list in Annex E
- "Company" historical data based on CAPAs
- MAUDE adverse events database
- Input from subject matter experts

1.1 Purpose

The purpose of this document is to provide a comprehensive list of all known and foreseeable hazards that are applicable to the products that are designed, developed or produced by "Companuy". Not every listed hazard is applicable to every product. In general, each product must be analyzed for the applicability of the entries in Table 1. Entries that are not applicable can be excluded in hazard analyses.

1.2 Definitions & Abbreviation

Term	Description
MAUDE	Manufacturer and User Facility Device Experience FDA's adverse events database www.accessdata.fda.gov/scripts/cdrh/cfdocs/cfmaude/search.cfm
CAPA	Corrective and Preventive Actions

See [Glossary] for other definitions.

Example CHL – Copyright 2018 Bijan Elahi

BXM	Clinical Hazards List	Doc # 12342 Revision 1.0 Page 2 of 2

2 CLINICAL HAZARDS LIST

Table 1 - "Company"Clinical Hazards List

ID	Hazard
Haz.01	No therapy
Haz.02	Inadequate therapy
Haz.03	Inappropriate shock
Haz.04	Hot surfaces
Haz.05	Sparks (pad to skin)
Haz.06	Ionizing radiation
Haz.07	Sharps
Haz.08	Pinch points
Haz.09	Current leakage
Haz.10	Vibration

3 REFERENCES

Reference	Document number	Title
[Glossary]	xxxxxx	QMS Glossary

4 REVISION HISTORY

Revision	Author	CR	Description of changes
1.0	John Adams	N/A	First approved version

Example CHL – Copyright 2018 Bijan Elahi

C.6 HARMS ASSESSMENT LIST

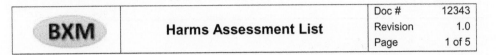

BXM	Harms Assessment List	Doc # 12343 Revision 1.0 Page 1 of 5

TABLE OF CONTENTS

Example HAL – Copyright 2018 Bijan Elahi

BXM	Harms Assessment List	Doc # 12343 Revision 1.0 Page 2 of 5

1 DOCUMENT INTRODUCTION

1.1 Purpose

This document lists the standardized set of potential harms related to the products that are designed, developed or produced by "Company". Additionally, for each listed harm the probability of possible outcomes are identified in 5 classes of: catastrophic, critical, serious, minor, and negligible.

The information provided can be used to compute the risks associated with relevant hazardous situations that "Company" products can present.

1.2 Abstract

The statistics in this Harms Assessment List (HAL) for defibrillation harms are based on the analysis of data in 14 published papers (see Appendix A for the raw data). In total 2,477 patients' data for 3,011 interventions (shocks) were taken into consideration.

The statistics for radiation therpay harms is based on analysis of data in 11 published papers (see Appendix B for the raw data). In total 733 patients' data for 1,386 interventions were taken into consideration.

1.3 Background

ISO 14971 defines risk as the product of probability of occurrence of a hazardous situation (P1), and the severity of the ensuing harm. A harm can affect a patient to different degrees. In this document five classes of harm severity are envisioned: Catastrophic, Critical, Serious, Minor and Negligible. See section 2 for the definitions of each harm class.

How to interpret the numbers in the HAL in section 2 – The risk equation is R = P1 x P2. Where P1 is the probability of occurrence of the hazardous situation. P2 is the probability of occurrence of harm in the various severity classes. This document provides the five P2 numbers for eah harm category. The reader should interpret this as: assuming the patient has been exposed to the hazard (P1 =100%), what are the chances of e.g., a catastrophic harm ($P2_{Catastrophic}$), or a critical harm ($P2_{Critical}$), etc. The P2 numbers are inclusive of normal countermeasure. For example in the case of a burn, in most cases

Example HAL – Copyright 2018 Bijan Elahi

BXM	Harms Assessment List	Doc # 12343 Revision 1.0 Page 3 of 5

medical care is exercised. And, in some cases medical care is not exercised. The P2 numbers account for both care, and no-care possibilities.

The Harm ID's are assigned as unique and permanent numbers. There is no special order in which the harms are cited in the HAL.

1.4 Definitions & Abbreviation

See [GLOSSARY] for a list of definitions and abbreviations.

1.5 References

Reference	Identification	Title / additional remarks
[GLOSSARY]	xxxxxx	QMS Glossary for "Company"

BXM	Harms Assessment List	Doc # 12343 Revision 1.0 Page 4 of 5

2 HARMS ASSESSMENT LIST

ID	Harm Category	Catastrophic	Critical	Serious	Minor	Negligible	Totals
Defibrillator Harms							
Harm.1	Burns (thermal)	0.0%	1.0%	70.0%	20.0%	9.0%	100.0%
Harm.4	Persistent fibrillation	85.0%	10.0%	5.0%	0.0%	0.0%	100.0%
Harm.9	Pain from therapeutic electric shock	0.0%	0.0%	10.0%	90.0%	0.0%	100.0%
Radiation Therapy Harms							
Harm.2	Burns (radiation)	5.0%	10.0%	80.0%	5.0%	0.0%	100.0%
Harm.5	Cell necrosis	0.1%	5.0%	80.0%	10.0%	4.9%	100.0%
Harm.7	Skin damage (blistering, peeling, dryness, ...)	0.0%	1.0%	85.0%	10.0%	4.0%	100.0%
Harm.8	Fatigue	0.0%	1.0%	40.0%	52.0%	7.0%	100.0%
Harm.3	Nausea	0.0%	0.0%	67.0%	22.0%	11.0%	100.0%
Common Harms							
Harm.6	Laceration	0.0%	2.0%	90.0%	7.0%	1.0%	100.0%
Harm.10	Mechanical harms (pinch, bump,..)	0.0%	0.0%	19.0%	76.0%	5.0%	100.0%
Harm.11	Electric shock	0.8%	4.0%	11.1%	4.1%	80.0%	100.0%

Example HAL – Copyright 2018 Bijan Elahi

BXM	Harms Assessment List	Doc # 12343 Revision 1.0 Page 5 of 5

Class	Definition
Catastrophic	Patient or operator death
Critical	Permanent impairment or life-threatening injury
Serious	Injury or impairment that requires professional medical intervention
Minor	Temporary injury or impairment not requiring professional medical intervention
Negligible	Inconvenience, or temporary discomfort

APPENDIX A – DATA FOR DEFIBRILLATION HARMS

[Raw data for defibrillation harms would be inserted here]

APPENDIX B – DATA FOR RADIATION THERAPY HARMS

[Raw data for radiation therapy harms would be inserted here]

REVISION HISTORY

Revision	Author	CR	Description of changes
1.0	John Adams	N/A	First approved version

Example HAL – Copyright 2018 Bijan Elahi

C.7 PRELIMINARY HAZARD ANALYSIS

BXM	PHA Vivio AED	Doc #	12341
		Revision	1.0
		Page	1 of 22

Table of Contents

Example PHA – Copyright 2018 Bijan Elahi

BXM	PHA Vivio AED	Doc # 12341 Revision 1.0 Page 2 of 22

1 GENERAL

In this document, the words "System", and "Vivio AED" are used synonymously.

1.1 Purpose

This document captures the Preliminary Hazard Analysis (PHA) of the Vivio AED. The purpose of the PHA is to identify the hazards, hazardous situations and events that could cause harm due to the use and operation of the Vivio AED. The PHA is performed early in the design and development process when there is little information on design details of the System. The PHA can be used to make an early estimation on whether the device can be made to an acceptable level of safety, and also to predict the safety-critical aspect of the device design. This information can be used to guide the product development team and focus resources on the safety-critical areas.

The basis for the PHA is the contents of Customer Requirements Specification [CRS] and the Technical Concept [TC].

1.2 Scope

The scope of this analysis is the Vivio AED System, excluding the Pads, which are supplied by third party. However, the interface between the AED and the Pads is within scope.

1.3 Definitions & Abbreviations

Term	Description
PHA	Preliminary Hazard Analysis
RACT	Risk Assessment and Control Table
FTA	Fault Tree Analysis
AED	Automatic External Defibrillator
IFU	Information for Use
VF	Ventricular Fibrillation

See [Glossary] for other definitions.

BXM	PHA Vivio AED	Doc # 12341 Revision 1.0 Page 3 of 22

2 ANALYSIS METHOD

In this Preliminary Hazard Analysis, the following steps are taken:

1) Product characteristics that could impact safety are identified.
2) The [CHL] is evaluated and relevant hazards to Vivio AED are identified
3) A Top-Down analysis is performed.
4) From steps 1-3, the potential causes that could lead into the hazards are identified.

Detailed examination of sequences of events that could lead into potential hazards will be analyzed in the System Hazard Analysis later in the project.

3 SYSTEM OVERVIEW

3.1 System Description and Intended Use

For System Description and Intended Use, see [RMP] section 2.

4 SAFETY CHARACTERISTICS

Safety characteristics of Vivio AED are listed below:

- Diagnosis and treatment of ventricular fibrillation via delivery of a high-voltage transthoracic electric shock
- Vivio AED provides audio/visual annunciations to guide use actions
- Effective delivery of essential performance depends on user actions, which are influenced by user-interface design
- Vivio AED operates on a primary battery which will deplete over its normal life
- Vivio AED performance is strongly influenced by its software performance
- Use of functional and quality Pads are imperative for the successful essential performance

Example PHA – Copyright 2018 Bijan Elahi

BXM	PHA Vivio AED	Doc # 12341 Revision 1.0 Page 4 of 22

5 APPLICABLE HAZARDS FROM THE CLINICAL HAZARD LIST

The Clinical Hazards List [CHL] is a list of all known and foreseeable hazards that are related to the devices made by "Company". For each hazard in [CHL] an analysis was performed to assess whether or not that hazard is relevant for the Vivio AED System. The result of this analysis is given in the table below.

ID	Hazard	Applicable?	Rationale
Haz.01	No therapy	Yes	Vivio AED delivers life-saving therapy
Haz.02	Inappropriate shock	Yes	Vivio AED may misdiagnose the heart rhythm and deliver an inappropriate shock
Haz.03	Hot surfaces	Yes	Vivio AED has potential to cause thermal burns
Haz.04	Sparks (pad to skin)	Yes	Improperly attached Pads could cause current jumping the airgap between pad and skin and cause sparks
Haz.05	Ionizing radiation	No	Vivio AED does not generate ionizing radiation
Haz.06	Sharps	Yes	Vivio AED has potential to present sharps
Haz.07	Pinch points	No	Vivio AED does not have potential for pinch points
Haz.08	Acid leakage	No	Vivio AED does not contain any acids
Haz.09	Current leakage	Yes	High voltage is present within Vivio
Haz.10	Vibration	No	Vivio AED does not have the potential to cause vibration hazards

6 TOP-DOWN ANALYSIS

A fault tree analysis of Vivio AED was performed to identify the potential pathways to System hazards. The top undesired event in each fault tree is one of the applicable hazards of Vivio AED. In each fault tree probability numbers were assigned to the basic events and thereby the probability of the top undesired event was derived. (Note – in this PHA the basic event probabilities are hypothetical, and assigned only as examples.)

Some of the basic events were not further developed. In some cases, it is because an element is out of scope, e.g. the Pads. In other cases, it is not possible to develop them further until more is known about the design. In the final hazard analysis, the bottom-up Failure Modes and Effects Analyses (FMEA) will provide additional details which will be captured in the RACT.

Figure 1 depicts the fault tree for the 'No Therapy' hazard.

Example PHA – Copyright 2018 Bijan Elahi

BXM	PHA Vivio AED	Doc #	12341
		Revision	1.0
		Page	5 of 22

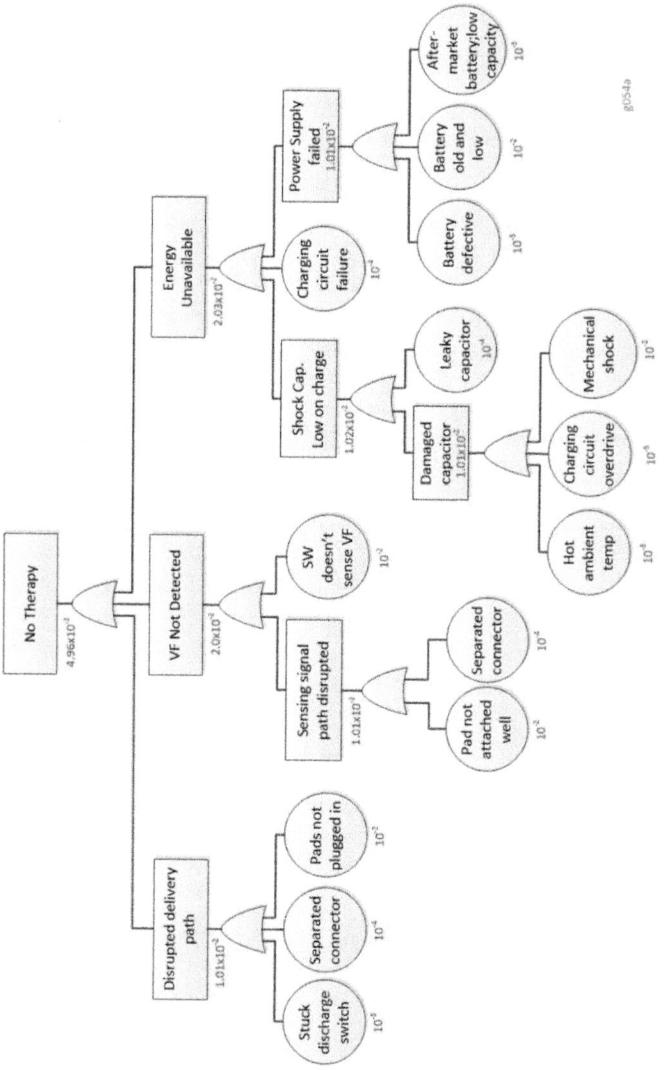

Figure 1- Vivio FTA - No Therapy

BXM	PHA Vivio AED	Doc # 12341
		Revision 1.0
		Page 6 of 22

Figure 2 combines the two hazards of 'Hot Surfaces' and 'Sparks' into one fault tree.

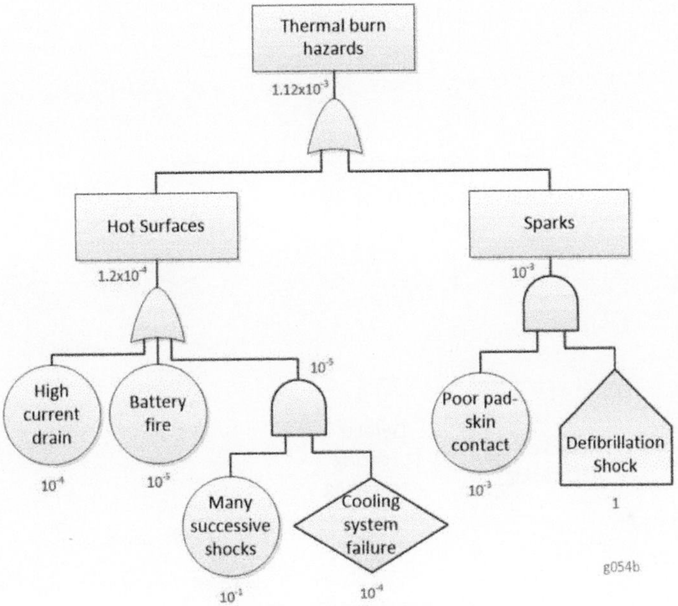

Figure 2 - Vivio FTA - Thermal Burn

Example PHA – Copyright 2018 Bijan Elahi

BXM	PHA Vivio AED	Doc #	12341
		Revision	1.0
		Page	7 of 22

Figure 3 depicts the fault tree for 'Sharps'.

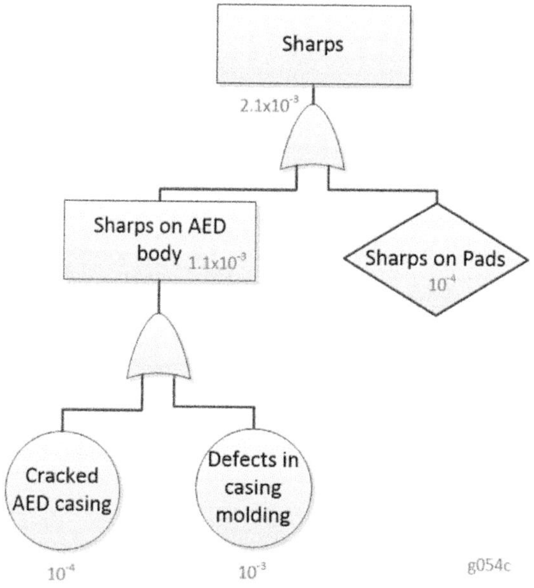

Figure 3 - Vivio FTA – Sharps

Figure 4 depicts the fault tree for 'Inappropriate Shock' and Figure 5 displays the fault tree for 'Current Leakage'.

Example PHA – Copyright 2018 Bijan Elahi

BXM	PHA Vivio AED	Doc # 12341 Revision 1.0 Page 8 of 22

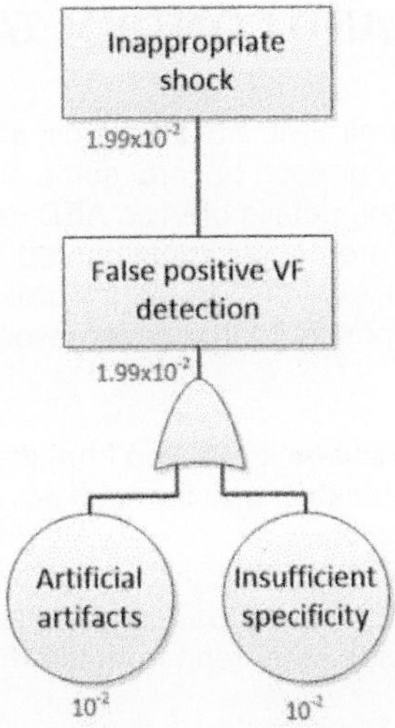

Figure 4 - Vivio FTA – Inappropriate shock

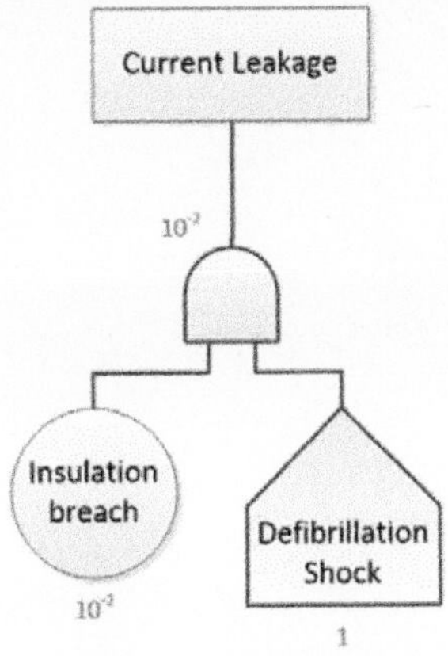

Figure 5 - Vivio FTA – Current Leakage

BXM	PHA Vivio AED	Doc #	12341
		Revision	1.0
		Page	9 of 22

7 RISK ASSESSMENT AND CONTROL TABLE (RACT)

The PHA RACT captures the overall Vivio AED hazards, their causes, their corresponding hazardous situations and harms. For each hazard, some risk controls are foreseen. At this early stage of product development, details of Vivio AED design are not yet available. This implies that the risk controls are not yet implemented. The P1 values are estimated to be inclusive of the foreseen risk controls. In other words, P1 is an estimate of the occurrence of the hazard, and exposure to that hazard while the foreseen risk controls are in place.

To calculate the risks, P2 numbers were looked up from the Harms Assessment List [HAL]. Each harm has five P2 estimates, one for each severity class of: Catastrophic, Critical, Serious, Minor, and Negligible.

The residual risks for each Hazard, Hazardous Situation, and overall residual risk for Vivio AED were computed. Risk acceptability criteria from the [RMP] are utilized to evaluate risks in each severity class.

These risk estimations will be updated in the final Hazard Analysis Report.

Example PHA – Copyright 2018 Bijan Elahi

BXM	PHA Vivio AED	Doc # 12341 Revision 1.0 Page 10 of 22

8 SOFTWARE SAFETY CLASSIFICATION

Based on the product description from [CRS] and the top-down analysis in section 6 above, it is concluded that the Software Safety Classification of Vivio AED according to IEC 62304 is Class C.

9 CONCLUSION AND RECOMMENDATIONS

The preliminary estimation of Vivio AED risks indicates that there is a potential for the device risks to exceed acceptable levels in the Catastrophic, Serious and Minor categories of harm. It is advised that the product development team investigate the potential for reduction of these risks to acceptable levels.

Additionally, the preliminary hazard analysis shows that the two areas of: VF detection software, and user interface design have high safety impacts and are worthy of extra attention in the product development process.

Example PHA – Copyright 2018 Bijan Elahi

BXM	PHA Vivio AED	Doc # 12341 Revision 1.0 Page 11 of 22

10 REFERENCES

Identification	Document number	Title
[CHL]	12342	Clinical Hazards List
[CRS]	xxxxxx	Customer Requirement Specifications
[Glossary]	xxxxxx	QMS Glossary
[HAL]	12343	Harms Assessment List
[RMP]	12340	RMP Vivio AED
[TC]	xxxxxx	Technical Concept

11 REVISION HISTORY

Revision	Author	Description of changes
1.0	John Adams	Initial version

Example PHA – Copyright 2018 Bijan Elahi

BXM	PHA Vivio AED	Doc # 12341 Revision 1.0 Page 12 of 22

APPENDIX A – PRODUCT CHARACTERIZATION

In this appendix, the results of evaluation of Vivio AED with respect to safety characteristics are captured. The questions in Annex C of ISO 14971 were used to guide this evaluation.

Question	Remarks
1. What is the intended use and how is the medical device to be used?	AED. Intended for use by minimally trained individuals on adults and children over 25 kg
2. Is the medical device intended to be implanted?	No
3. Is the medical device intended to be in contact with the patient or other persons?	Vivio AED does not contact patients. But third-party supplied defibrillation pads do contact skin surface
4. What materials or components are utilized in the medical device or are used with, or are in contact with, the medical device?	Vivio AED does not contact patient
5. Is energy delivered to or extracted from the patient?	Yes
6. Are substances delivered to or extracted from the patient?	No
7. Are biological materials processed by the medical device for subsequent re-use, transfusion or transplantation?	No
8. Is the medical device supplied sterile or intended to be sterilized by the user, or are other microbiological controls applicable?	No
9. Is the medical device intended to be routinely cleaned and disinfected by the user?	Minor cleaning with damp cloth is sufficient

Example PHA – Copyright 2018 Bijan Elahi

BXM	PHA Vivio AED	Doc # 12341 Revision 1.0 Page 13 of 22

Question	Remarks
10. Is the medical device intended to modify the patient environment?	No
11. Are measurements taken?	Yes
12. Is the medical device interpretative?	No
13. Is the medical device intended for use in conjunction with other medical devices, medicines or other medical technologies?	Yes. Third-party supplied defibrillation pads.
14. Are there unwanted outputs of energy or substances?	No
15. Is the medical device susceptible to environmental influences?	Yes. Vivio AED is not water proof, and should be operated within temperature range of: 0° - 50° C, and altitude of 0 – 4,500 m above sea level
16. Does the medical device influence the environment?	No
17. Are there essential consumables or accessories associated with the medical device?	Yes. Third-party supplied defibrillation pads.
18. Is maintenance or calibration necessary?	No calibration. But minor cleaning and, battery replacement is necessary.
19. Does the medical device contain software?	Yes
20. Does the medical device have a restricted shelf-life?	No
21. Are there any delayed or long-term use effects?	No
22. To what mechanical forces will the medical device be subjected?	Normal handling and potentially drops from up to 1 m.
23. What determines the lifetime of the medical device?	Delivery of shocks depletes the battery, which is normal.
24. Is the medical device intended for single use?	No

Example PHA – Copyright 2018 Bijan Elahi

BXM	PHA Vivio AED	Doc # 12341 Revision 1.0 Page 14 of 22

Question	Remarks
25. Is safe decommissioning or disposal of the medical device necessary?	Disposal according to local electronic waste rules should be followed.
26. Does installation or use of the medical device require special training or special skills?	No. The device is designed for use by minimally trained individuals. Just following the IFU is sufficient.
27. How will information for safe use be provided?	Color LCD will provide graphic, visual guidance. Also, audio guidance will be provided in local language. The IFU also provides the same in print format.
28. Will new manufacturing processes need to be established or introduced?	No
29. Is successful application of the medical device critically dependent on human factors such as the user interface?	Yes
29.1 Can the user interface design features contribute to use error?	Yes
29.2 Is the medical device used in an environment where distractions can cause use error?	Yes
29.3 Does the medical device have connecting parts or accessories?	Yes
29.4 Does the medical device have a control interface?	Yes. A very simple two-button operation
29.5 Does the medical device display information?	Yes
29.6 Is the medical device controlled by a menu?	No
29.7 Will the medical device be used by persons with special needs?	No. Although it is possible that a person with special-needs attempts to use the device.
29.8 Can the user interface be used to initiate user actions?	Yes

Example PHA – Copyright 2018 Bijan Elahi

BXM	PHA Vivio AED	Doc # 12341 Revision 1.0 Page 15 of 22

Question	Remarks
30. Does the medical device use an alarm system?	Yes
31. In what way(s) might the medical device be deliberately misused?	The device may be used on small children or infants
32. Does the medical device hold data critical to patient care?	No
33. Is the medical device intended to be mobile or portable?	Yes
34. Does the use of the medical device depend on essential performance?	Yes

Example PHA – Copyright 2018 Bijan Elahi

BXM	PHA Vivio AED	Doc # 12341 Revision 1.0 Page 16 of 22

APPENDIX B – PHA RACT

The RACT tables in the following pages are split in the opposing pages to allow larger font and better readability.

Example PHA – Copyright 2018 Bijan Elahi

BXM	PHA Vivio AED	Doc # 12341
		Revision 1.0
		Page 17 of 22

BXM

PHA RACT - Risk per Hazard
Vivio EAD

ID	Initial cause of hazard	Sequence of Events	Hazard	Hazardous Situations	Risk Controls SD	PM	IS
1	SW doesn't sense VF	SW doesn't sense VF → VF not detected → Therapy not delivered	Haz.01 No Therapy	Fibrillating patient does not receive therapy	Sensitive SW algorithm	N/A	N/A
3	Battery defective	Battery defective → Power supply failed → Energy unavailable → Therapy not delivered	Haz.01 No Therapy	Fibrillating patient does not receive therapy	Initial battery health check upon installation	Periodic battery checks	recomm. to buy quality batteries
4	Circuit failure	Undetermined electronic circuit failure → Therapy not delivered	Haz.01 No Therapy	Fibrillating patient does not receive therapy	electronic provisions to handle circuit failures	N/A	N/A
9	Failed UI	Undetermined causes → UI defect → user does not deliver tharapy to patient	Haz.01 No Therapy	Fibrillating patient does not receive therapy	proper UI design to prevent use-failures	protective measures in UI to aid user	alarms/alerts to help user take corrective actions
12	User unable to clean and prepare skin for defib pad application	Unable to clean and prepare skin → defib pads partially adhere to skin → Full shock energy is not delivered to patient	Haz.01 No Therapy	Fibrillating patient receives inadequate shock energy for defibrillation	N/A	Razor and alcohol wipes supplied in the Vivio outer case	Audio/visual and IFU instruct to prepare skin
13	Improperly place the pads on patient's chest	Perception error - cannot see visual information	Haz.01 No Therapy	Fibrillating patient receives inadequate shock energy for defibrillation	N/A	N/A	* Large-print diagram on the pad pouch * Animated graphics on LCD display * Audio guidance
8	SW mis-detects VF	Insufficient specificity → False positive VF detection → Inappropriate shock	Haz.02 Inappropriate shock	Patient without VF receives defibrillation shock	SW algorithm with high specificity	N/A	N/A
14	Artifacts	Artifacts confuse SW → False positive VF detection	Haz.02 Inappropriate shock	Patient without VF receives defibrillation shock	SW algorithm with high specificity	N/A	Instructions for proper use of the device
5	Circuit failure	Loss of feedback loop to μC → Charging circuit charges indefinitely → Increased internal temperature	Haz.03 Hot surfaces	User or bystander comes in contact with hot device	Thermistor feedback interlock to prevent runaway charging	Plastic casing is a poor heat conductor	N/A
6	Leave hair or dirt on patient's skin	Unable to clean and prepare skin → defib pads partially adhere to skin → sparks between pads and skin	Haz.04 Sparks (pad to skin)	Patient exposed to high voltage spark to skin	SW monitors EKG signal quality and warns of bad contact	Razor and alcohol wipes supplied in the Vivio outer case	audio visual guidance on UI
7	Impact to casing	Impact to casing → cracked AED casing → sharps on AED body	Haz.06 Sharps	User or bystander comes in contact with a sharp edge/point	Casing made of ABS to withstand high impact	Soft grip surface to prevent dropping	Caution in IFU to not treat the AED roughly
15	Insulation breach	Erosion of insulation → high voltage lines become exposed	Haz.09 Current Leakage	User exposed to high voltage	Use of high integrity insulation	Packaging protects pad wires	Warning to user to stand clear

Example PHA RACT - Copyright 2018 Bijan Elahi

Example PHA – Copyright 2018 Bijan Elahi

BXM	PHA Vivio AED	Doc # 12341 Revision 1.0 Page 18 of 22

Doc # 12347
Revision 1.0

P1	Harm	P2					Risk				
		Cat	Crit	Ser	Minr	Negl	Cat	Crit	Ser	Minr	Negl
0	Harm.4 Persistent fibrillation	0.85	0.1	0.05	0	0	8.5E-06	1.0E-06	5.0E-07	0.0E+00	0.0E+00
0	Harm.4 Persistent fibrillation	0.85	0.1	0.05	0	0	8.5E-06	1.0E-06	5.0E-07	0.0E+00	0.0E+00
0	Harm.4 Persistent fibrillation	0.85	0.1	0.05	0	0	8.5E-06	1.0E-06	5.0E-07	0.0E+00	0.0E+00
0	Harm.4 Persistent fibrillation	0.85	0.1	0.05	0	0	8.5E-06	1.0E-06	5.0E-07	0.0E+00	0.0E+00
0	Harm.4 Persistent fibrillation	0.85	0.1	0.05	0	0	4.3E-05	5.0E-06	2.5E-06	0.0E+00	0.0E+00
0	Harm.4 Persistent fibrillation	0.85	0.1	0.05	0	0	4.3E-05	5.0E-06	2.5E-06	0.0E+00	0.0E+00
						Haz.01 No Therapy →	1.10E-04	1.40E-05	7.00E-06	0.00E+00	0.00E+00
####	Harm.9 Pain from electric shock	0	0	0.1	0.9	0	0.0E+00	0.0E+00	1.0E-04	9.0E-04	0.0E+00
####	Harm.9 Pain from electric shock	0	0	0.1	0.9	0	0.0E+00	0.0E+00	1.0E-04	9.0E-04	0.0E+00
						Haz.02 Inappropriate Shock →	0.0E+00	0.0E+00	2.0E-04	1.8E-03	0.0E+00
0	Harm.1 Burns (thermal)	0	0.01	0.7	0.2	0.09	0.0E+00	1.0E-08	7.0E-07	2.0E-07	9.0E-08
0	Harm.1 Burns (thermal)	0	0.01	0.7	0.2	0.09	0.0E+00	1.0E-06	7.0E-05	2.0E-05	9.0E-06
1E-05	Harm.6 Laceration	0	0.02	0.9	0.07	0.01	0.0E+00	2.0E-07	9.0E-06	7.0E-07	1.0E-07
1E-06	Harm.11 Electric Shock	0.008	0.04	0.111	0.041	0.8	8.0E-09	4.0E-08	1.1E-07	4.1E-08	8.0E-07

Page 1 of 6

Example PHA – Copyright 2018 Bijan Elahi

BXM	PHA Vivio AED	Doc # 12341 Revision 1.0 Page 19 of 22

BXM

PHA RACT - Risk per Hazardous Situation Overall Vivio EAD

ID	Initial cause of hazard	Sequence of Events	Hazard	Hazardous Situations	Risk Controls SD	PM	IS
1	SW doesn't sense VF	SW doesn't sense VF → VF not detected → Therapy not delivered	Haz.01 No Therapy	Fibrillating patient does not receive therapy	Sensitive SW algorithm	N/A	N/A
3	Battery defective	Battery defective → Power supply failed → Energy unavailable → Therapy not delivered	Haz.01 No Therapy	Fibrillating patient does not receive therapy	Initial battery health check upon installation	Periodic battery checks	recomm. to buy quality batteries
4	Circuit failure	Undetermined electronic circuit failure → Therapy not delivered	Haz.01 No Therapy	Fibrillating patient does not receive therapy	electronic provisions to handle circuit failures	N/A	N/A
9	Failed UI	Undetermined causes → UI defect → user does not deliver tharapy to patient	Haz.01 No Therapy	Fibrillating patient does not receive therapy	proper UI design to prevent use-failures	protective measures in UI to aid user	alarms/alerts to help user take corrective actions
12	User unable to clean and prepare skin for defib pad application	Unable to clean and prepare skin → defib pads partially adhere to skin → Full shock energy is not delivered to patient	Haz.01 No Therapy	Fibrillating patient receives inadequate shock energy for defibrillation	N/A	Razor and alcohol wipes supplied in the Vivio outer case	Audio/visual and IFU instruct to prepare skin
13	Improperly place the pads on patient's chest	Perception error - cannot see visual information	Haz.01 No Therapy	Fibrillating patient receives inadequate shock energy for defibrillation	N/A	N/A	* Large-print diagram on the pad pouch * Animated graphics on LCD display * Audio guidance
				→			
8	SW mis-detects VF	Insufficient specificity → False positive VF detection → Inappropriate shock	Haz.02 Inappropriate shock	Patient without VF receives defibrillation shock	SW algorithm with high specificity	N/A	N/A
14	Artifacts	Artifacts confuse SW → False positive VF detection	Haz.02 Inappropriate shock	Patient without VF receives defibrillation shock	SW algorithm with high specificity	N/A	Instructions for proper use of the device
				→			
5	Circuit failure	Loss of feedback loop to μC → Charging circuit charges indefinitely → Increased internal temperature	Haz.03 Hot surfaces	User or bystander comes in contact with hot device	Thermistor feedback interlock to prevent runaway charging	Plastic casing is a poor heat conductor	N/A
6	Leave hair or dirt on patient's skin	Unable to clean and prepare skin → defib pads partially adhere to skin → sparks between pads and skin	Haz.04 Sparks (pad to skin)	Patient exposed to high voltage spark to skin	SW monitors EKG signal quality and warns of bad contact	Razor and alcohol wipes supplied in the Vivio outer case	audio visual guidance on UI
7	Impact to casing	Impact to casing → cracked AED casing → sharps on AED body	Haz.06 Sharps	User or bystander comes in contact with a sharp edge/point	Casing made of ABS to withstand high impact	Soft grip surface to prevent dropping	Caution in IFU to not treat the AED roughly
15	Insulation breach	Erosion of insulation → high voltage lines become exposed	Haz.09 Current Leakage	User exposed to high voltage	Use of high integrity insulation	Packaging protects pad wires	Warning to user to stand clear

Example PHA RACT - Copyright 2018 Bijan Elahi
Example PHA – Copyright 2018 Bijan Elahi

BXM	PHA Vivio AED	Doc # 12341 Revision 1.0 Page 20 of 22

Doc # 12347
Revision 1.0

P1	Harm	P2					Risk				
		Cat	Crit	Ser	Minr	Negl	Cat	Crit	Ser	Minr	Negl
1.0E-05	Harm.4 Persistent fibrillation	0.85	0.1	0.05	0	0	8.5E-06	1.0E-06	5.0E-07	0.0E+00	0.0E+00
1.0E-05	Harm.4 Persistent fibrillation	0.85	0.1	0.05	0	0	8.5E-06	1.0E-06	5.0E-07	0.0E+00	0.0E+00
1.0E-05	Harm.4 Persistent fibrillation	0.85	0.1	0.05	0	0	8.5E-06	1.0E-06	5.0E-07	0.0E+00	0.0E+00
1.0E-05	Harm.4 Persistent fibrillation	0.85	0.1	0.05	0	0	8.5E-06	1.0E-06	5.0E-07	0.0E+00	0.0E+00
5.0E-05	Harm.4 Persistent fibrillation	0.85	0.1	0.05	0	0	4.3E-05	5.0E-06	2.5E-06	0.0E+00	0.0E+00
5.0E-05	Harm.4 Persistent fibrillation	0.85	0.1	0.05	0	0	4.3E-05	5.0E-06	2.5E-06	0.0E+00	0.0E+00
Fibrillating patient does not receive adequate therapy →							1.19E-04	1.40E-05	7.00E-06	0.00E+00	0.00E+00
1.0E-03	Harm.9 Pain from electric shock	0	0	0.1	0.9	0	0.0E+00	0.0E+00	1.0E-04	9.0E-04	0.0E+00
1.0E-03	Harm.9 Pain from electric shock	0	0	0.1	0.9	0	0.0E+00	0.0E+00	1.0E-04	9.0E-04	0.0E+00
Patient without VF receives defibrillation shock →							0.0E+00	0.0E+00	2.0E-04	1.8E-03	0.0E+00
1.0E-06	Harm.1 Burns (thermal)	0	0.01	0.7	0.2	0.09	0.0E+00	1.0E-08	7.0E-07	2.0E-07	9.0E-08
1.0E-04	Harm.1 Burns (thermal)	0	0.01	0.7	0.2	0.09	0.0E+00	1.0E-06	7.0E-05	2.0E-05	9.0E-06
1.0E-05	Harm.6 Laceration	0	0.02	0.9	0.07	0.01	0.0E+00	2.0E-07	9.0E-06	7.0E-07	1.0E-07
1.0E-06	Harm.11 Electric Shock	0.008	0.04	0.111	0.041	0.8	8.0E-09	4.0E-08	1.1E-07	4.1E-08	8.0E-07
			Overall Residual Risk →				1.2E-04	1.5E-05	2.9E-04	1.8E-03	1.0E-05

Page 2 of 6

Example PHA – Copyright 2018 Bijan Elahi

BXM	PHA Vivio AED	Doc # 12341
		Revision 1.0
		Page 21 of 22

BXM

PHA RACT - Vivio AED
Acceptable Risk Limits

Doc # 12347
Revision 1.0

The following risk limits are derived from a survey of published data on rates of occurrence of harm to patients from the use of AEDs. This is construed as the state-of-the-art for acceptable risk levels for AED use.

R-Cat	R-Crit	R-Ser	R-Minr	R-Negl
7.1E-05	9.5E-05	1.3E-04	1.5E-03	1.0E-02

Example PHA RACT - Copyright 2018 Bijan Elahi

Example PHA – Copyright 2018 Bijan Elahi

BXM	PHA Vivio AED	Doc # 12341 Revision 1.0 Page 22 of 22

BXM

PHA RACT
Vivio AED

Doc # 12347
Revision 1.0

Abbreviations

Term	Definition
Cat	Catastrophic
Crit	Critical
ID	Identification
IS	Information for Safety
Minr	Minor
Negl	Negligible
PM	Protective Measure
RACT	Risk Assessment and Control Table
SD	Safe by Design
Ser	Serious

Revision History

Revision	Author	CR	Description of Change
1.0	John Adams	N/A	First approved version

C.8 DESIGN FAILURE MODES AND EFFECTS ANALYSIS

BXM

DFMEA
Vivio AED

Doc # 12344
Revision 1.0

Scope

This DFMEA covers the Vivio AED design.

The scope of the analysis is bounded in the diagram below and encompases all the items within the analysis boundary.

Item Under Analysis: Vivio AED model 1234, version 1.1

Primary functions: Detect VF and deliver therapeutic shock

Secondary Functions: Monitor AED health and report any abnormalities
Monitor the connections of the defibrillation pads, both to Vivio and to patient's skin

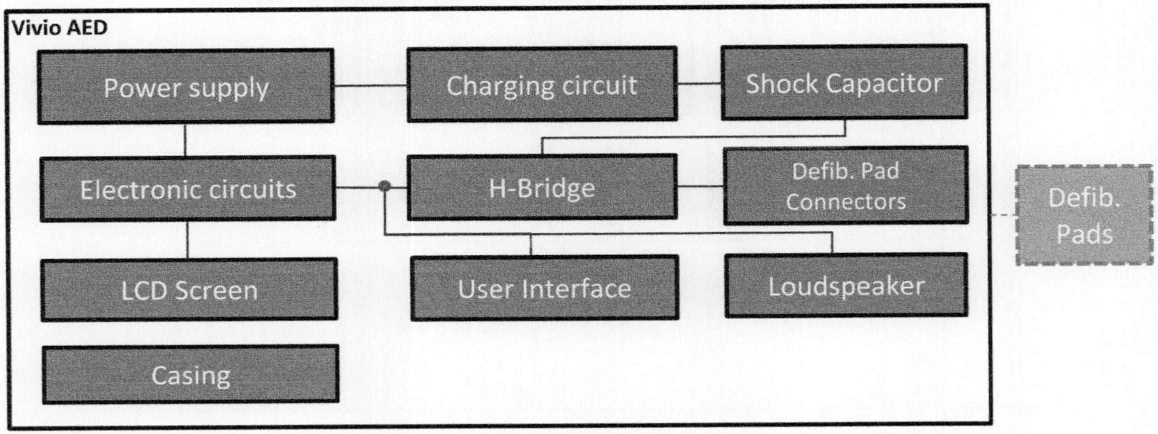

Copyright 2018 Bijan Elahi

BXM

DFMEA
Vivio AED

Doc # 12344
Revision 1.0

ITEM / FUNCTION				POTENTIAL FAILURE MODES & EFFECTS			
ID	**Source**	**Item**	**Function**	**Failure Mode**	**Causes /Mechanisms of Failure**	**Local Effects of Failure**	**End Effects of Failure**
1	DFMEA PS ID2	Power Supply	Provide power to all electronics	No power	Battery fails → Power supply failed → Energy unavailable	N/A	No therapy delivered
2	DFMEA PS ID6	Power Supply	Provide power to all electronics	Unsteady power (varying voltage)	Temperature sensitivity of battery → battery voltage fluctuates with temp.	inadequate voltage to internal electronics	Inadequate defib energy
3	DFMEA CC ID9	Charging Circuit	Charge the shock capacitor	No charging	Excess current → MOSFET failure → Inability to provide high voltage to shock capacitor	N/A	No therapy delivered
4	DFMEA CC ID7	Charging Circuit	Charge the shock capacitor	Inadequate charging	Over voltage → Rectifying diode failure → Current leakage → Inability to charge the shock cap. to target voltage	N/A	Inadequate defib energy
5	DFMEA CC ID10	Charging Circuit	Charge the shock capacitor	Runaway charging	Loss of feedback loop to µC → Charging circuit charges indefinitely → Increased internal temperature	Overheating internal electronics	Hot surfaces
6	N/A	Shock Capacitor	Accumulate energy for defibrillation	Capacitor fails short	Over voltage → dielectric breakdown → shorted capacitor	N/A	No therapy delivered
7	N/A	Shock Capacitor	Accumulate energy for defibrillation	Capacitor leaks	Corrosion of leads → leakage of current → loss of capacitor charge	N/A	Inadequate defib energy
8	DFMEA DS ID3	H-Bridge	Connect defib pads to shock capacitor, while simultaneously disconnecting the detection circuit	Does not connect defib pads to shock capacitor	Gate line opens → FET switch not activated → Shock not delivered	N/A	No therapy delivered
9	DFMEA DS ID5	H-Bridge	Connect defib pads to shock capacitor, while simultaneously disconnecting the detection circuit	Does not disconnect detection circuit from defib pads	High voltage on drain → electrical overstress on gate → FET failure → High voltage feedback into EKG circuits → destruction of sensing circuits	Sensing circuit destroyed by defib shock	Shock is delivered but sensing circuit is destroyed and future shock are not delivered
10	DFMEA DS ID6	H-Bridge	Connect defib pads to shock capacitor, while simultaneously disconnecting the detection circuit	Does not connect defib pads to detection circuit	Gate line opens → FET switch not activated → Sensing circuits are not connected to defib pads → VF not detected	N/A	No therapy delivered
11	N/A	Defib Pad Connector	Electrically connect defib pads with AED	No electrical contact	Pad connector plug slips out of socket → AED not connected to defib pads → shock not delivered	N/A	No therapy delivered

Example DFMEA - Copyright 2018 Bijan Elahi

BXM

DFMEA
Vivio AED

Doc # 12344
Revision 1.0

		INITIAL RATING					FINAL RATING				
Safety Impact?	**Existing Mitigations**	**Sev**	**Occ**	**Det**	**RPN (auto)**	**Additional Mitigations**	**Sev**	**Occ**	**Det**	**RPN (auto)**	**Remarks**
Y	N/A	5	2	1	10	* SW does battery health check upon installation * Periodic battery check & alarm if battery has failed	5	1	1	5	
N	N/A	5	3	4	60	Voltage regulator compensates for temperature variability	5	1	4	20	Safety Impact=N because end effect is eliminated
Y	N/A	5	2	3	30	Current limiter prevents over-current into MOSFET	5	1	3	15	
N	N/A	5	2	4	40	Voltage regulator prevents over voltage stressing of rectifyer diode.	5	1	4	20	Safety Impact=N because end effect is eliminated
Y	N/A	3	3	2	18	Thermistor feedback interlock to prevent runaway charging	3	2	2	12	
N	N/A	5	2	3	30	Voltage regulator prevents over voltage	5	1	3	15	Safety Impact=N because end effect is eliminated
Y	N/A	5	3	4	60	Capacitor leads are hermetically sealed to prevent corrosion	5	1	4	20	
Y	N/A	5	2	3	30	Redundant gate connection with welded wires.	5	1	3	15	
N	N/A	3	3	4	36	FET gates are protected against high voltage	3	1	4	12	
Y	N/A	5	2	3	30	Redundant gate connection with welded wires.	5	1	3	15	
Y	N/A	5	3	3	45	* 1 N spring force to retain plug * Positive click for haptic feedback upon connection	5	1	3	15	

Example DFMEA - Copyright 2018 Bijan Elahi

BXM

DFMEA
Vivio AED

Doc # 12344
Revision 1.0

ITEM / FUNCTION				POTENTIAL FAILURE MODES & EFFECTS			
ID	Source	Item	Function	Failure Mode	Causes /Mechanisms of Failure	Local Effects of Failure	End Effects of Failure
12	N/A	Defib Pad Connector	Electrically connect defib pads with AED	Intermittent electrical contact	Weak contact between plug and socket → intermittent electrical contact → noise on VF sensing → False positive VF detection	N/A	Inappropriate shock delivered
13	N/A	LCD Screen	Provide visual guidance to user	Screen failed in off mode	Backlight failure → screen goes dark → user cannot get visual cues	N/A	Inconvenience to user
14	N/A	LCD Screen	Provide visual guidance to user	Random/flashing pixels	Solder connection breaks → Abnormal display	N/A	Inconvenience to user
15	N/A	UI Buttons	Receive input from user	Stuck open/closed	UI button aging → button stuck in open/closed mode → shock not delivered		No therapy delivered
16	N/A	Speaker	Provide audio output	No sound	Amplifier gain too high → Too much power to loudspeaker → Voice coil open circuit failure → No acoustic output	N/A	No sound output
17	N/A	Speaker	Provide audio output	Garbled sound	Amplifier gain too high → Too much power to loudspeaker → blown voice coil → distorted output	N/A	Distorted sound output
18	N/A	Casing	Provide protection and containment of AED internals	Cracked casing	Impact to casing → cracked AED casing → sharps on AED body	N/A	Sharps

<This example DFMEA is truncated>

BXM

DFMEA
Vivio AED

Doc # 12344
Revision 1.0

		INITIAL RATING					FINAL RATING				
Safety Impact?	Existing Mitigations	Sev	Occ	Det	RPN (auto)	Additional Mitigations	Sev	Occ	Det	RPN (auto)	Remarks
Y	N/A	3	3	2	18	* 1 N spring force to retain plug	3	2	2	12	
N	Existing design uses a high reliability LCD screen. The AED will provide audio cues in addition to dedicated lights on the UI buttons. Shock is delivered as needed.	2	2	1	4	N/A	2	2	1	4	
N	Use welded contacts	2	2	2	8	N/A	2	1	2	4	
Y	* Use of high reliability switches * Periodic self check; AED warns of failure PRIOIR to use of the deivce	5	1	3	15	N/A	5	1	3	15	
N	N/A	3	3	2	18	Audio drivers designed to prevent overpowering the loudspeaker	3	2	2	12	
Y	N/A	4	2	1	8	Audio drivers designed to prevent overpowering the loudspeaker	4	1	1	4	
Y	* Casing made of ABS to withstand high impact * Soft grip surface to prevent dropping	3	2	2	12	N/A	3	2	2	12	

Example DFMEA - Copyright 2018 Bijan Elahi

Severity Criteria (Sev)		
Rank	**Severity Descriptions - No Safety Impact**	**Severity Description -- Safety Impact**
5	**Catastrophic:** Described failure mode will cause immediate failure of the Subject. (Total loss of all functions – primary and secondary)	**Catastrophic** – Impact of the end-effect at the System level can be death
4	**Critical:** Described failure mode will severely impact Subject functionality | Complete loss of primary functions	**Critical** – Impact of the end-effect at the System level can be permanent impairment or life-threatening injury
3	**Serious:** Described failure mode will reduce Subject functionality. (Partial loss of primary functions | Complete loss of secondary functions)	**Serious** – Impact of the end-effect at the System level can be injury or impairment that requires professional medical intervention
2	**Minor:** Described failure mode will have temporal or self-restoring impact on functionality | partial loss of secondary functions	**Minor** – Impact of the end-effect at the System level can be temporary injury or impairment that does not require professional medical intervention
1	**None:** Described component failure will have no impact on functionality	**Negligible** – Impact of the end-effect at the System level can be at most an inconvenience, or temporary discomfort

RPN	Action
53-125	**Level 3** - Reduce RPN through failure compensating provisions.
13-52	**Level 2** - If Safety Impact is Y, reduce RPN to as low as possible. If Safety Impact is N, reduce RPN if feasible.
1-12	**Level 1** - If Safety Impact is Y, reduce RPN to as low as possible. If Safety Impact is N, further RPN reduction is not required.

Probability of Occurrence Criteria (Occ)			
Category	**Rank**	**Qualitative Criteria**	**Quantitative Criteria**
Frequent	5	The occurrence is frequent. Failure may be almost certain or constant failure.	$\geq 10^{-3}$
Probable	4	The occurrence is probable. Failure may be likely | repeated failures are expected.	$< 10^{-3}$ and $\geq 10^{-4}$
Occasional	3	The occurrence is occasional. Failures may occur at infrequent intervals.	$< 10^{-4}$ and $\geq 10^{-5}$
Remote	2	The occurrence is remote. Failures are seldom expected to occur.	$< 10^{-5}$ and $\geq 10^{-6}$
Improbable	1	The occurrence is improbable. The failure is not expected to occur.	$< 10^{-6}$

Detection Criteria (Det)			
Category	**Rank**	**Qualitative Criteria**	**Quantitative Criteria**
Undetectable	5	No detection opportunity | No means for detection | Countermeasures not possible	$< 10^{-3}$
Low	4	Opportunity for detection is low | Countermeasures are unlikely	$< 10^{-2}$ and $\geq 10^{-3}$
Moderate	3	Opportunity for detection is moderate | Countermeasures are probable	$< 10^{-1}$ and $\geq 10^{-2}$
High	2	Opportunity for detection is high | Countermeasures are likely	$< 9 \times 10^{-1}$ and $\geq 10^{-1}$
Almost Certain	1	Opportunity for detection is almost certain | Countermeasures are certain	$\geq 9 \times 10^{-1}$

Example DFMEA - Copyright 2018 Bijan Elahi

BXM

DFMEA
Vivio AED

Doc # 12344
Revision 1.0

Revision History

Revision	Author	CR	Description of Change
1.0	John Adams	N/A	First approved version

BXM

DFMEA - Vivio AED
Log of Working Sessions

Doc # 12344
Revision 1.0

Date	Participants
2017/1/10	John Adams, James Polk, Abe Lincoln, John Kennedy
2017/1/25	John Adams, James Polk, Abe Lincoln, John Kennedy
2017/2/16	John Adams, James Polk, Abe Lincoln, John Kennedy, Sonia Sotomayor

Example DFMEA - Copyright 2017 Bijan Elahi

C.9 PROCESS FAILURE MODES AND EFFECT ANALYSIS

BXM

PFMEA
Vivio AED

Doc # 12345
Revision 1.0

Scope

This PFMEA covers the manufacturing process for Vivio AED.

The scope of the analysis is bounded in the diagram below and encompases all the items within the analysis boundary.

Process Under Analysis: Manufacturing process MFP.v.03, for Vivio AED model 1234, version 1.1

Primary functions: Detect VF and deliver therapeutic shock

Secondary Functions: Monitor AED health and report any abnormalities
Monitor the connections of the defibrillation pads, both to Vivio and to patient's skin

Example PFMEA - Copyright 2018 Bijan Elahi

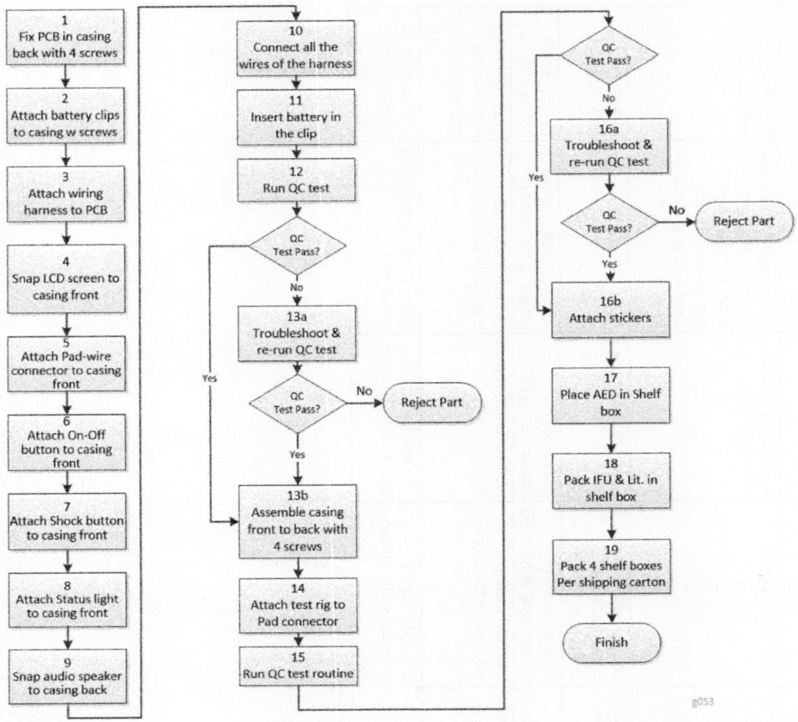

Example PFMEA - Copyright 2018 Bijan Elahi

BXM

PFMEA
Vivio AED

Doc # 12345
Revision 1.0

ITEM / FUNCTION			POTENTIAL FAILURE MODES & EFFE		
ID	**Process Step**	**Process Step Function**	**Failure Mode**	**Causes /Mechanisms of Failure**	**Local Effects of Failure**
Task 1 - Assemble Vivio					
1	Fix PCB in casing back with 4 screws	Afix PCB to casing	One or more screws missing	Assembly worker forgets to use all 4 screws	N/A
2	Fix PCB in casing back with 4 screws	Afix PCB to casing	Afix PCB in wrong orientaion	Symetrical PCB design	PCB connectors in wrong orientation
3	Fix PCB in casing back with 4 screws	Afix PCB to casing	ESD shock to PCB while installing	Worker/workstation are not properly grounded → Worker handles PCB and caused an ESD discharge and damage to PCB	FET switch failure
4	Attach battery clips to casing w screws	Secure battery and make electrical connection	Forget one or more screws	Error in assembly	Battery cannot be installed
5	Attach wiring harness to PCB	Enable electrical connections within AED	Incomplete connection to PCB	Worker unclear when proper connection is	Device would not function
6	Snap LCD screen to casing front	Get LCD display mounted in the proper place	LCD mounted upside down	Symmetrical LCD assembly design	LCD connector in the wrong orientation
7	Snap LCD screen to casing front	Get LCD display mounted in the proper place	Too much force is applied to LCD assy.	Snap force for mounting is too high	LCD solder joints are damaged

<This example PFMEA is truncated>

Example PFMEA - Copyright 2018 Bijan Elahi

BXM

PFMEA
Vivio AED

Doc # 12345
Revision 1.0

CTS			INITIAL RATING					FINAL RATING				
End Effects of Failure	Safety Impact?	Existing Mitigations	Sev	Occ	Det	RPN (auto)	Additional Mitigations	Sev	Occ	Det	RPN (auto)	Remarks
PCB could be loose and dislodge	N	N/A	2	3	3	18	* 4 screws in kit tray; obvious unused screw * Machine vision detection	2	1	2	4	
Assembly cannot be completed	N	N/A	2	3	3	18	* asymmetric screw pattern	2	1	1	2	
No therapy delivered	Y	N/A	5	3	5	75	* Require ESD wrist strap * Grounded workstation * Ion fans on work area	5	1	5	25	
No therapy delivered	N	N/A	2	3	5	30	QC test would catch an unpowered AED	2	1	5	10	
No therapy delivered	N	N/A	2	3	3	18	Design for snap-click on proper connection	2	1	3	6	
Assembly cannot be completed	N	N/A	2	3	2	12	Design feature in casing prevents insertion of LCD in wrong orientation	2	1	2	4	
Erratic display	N	N/A	4	3	3	36	Snap force is designed to 3 N	4	1	3	12	

Example PFMEA - Copyright 2018 Bijan Elahi

Severity Criteria (Sev)		
Rank	**Severity Descriptions – Non-Safety**	**Severity Description – Safety Impact**
5	Failure to meet Regulatory requirements | Total loss of all functions – primary and secondary | >70% of the production has to be scrapped	**Catastrophic** – Impact of the end-effect at the System level can be death
4	Loss or degradation of primary functions | Failure to meet product specification | Scrapping of 50-70% of the production	**Critical** – Impact of the end-effect at the System level can be permanent impairment or life-threatening injury
3	Loss or degradation of secondary functions | Reduced reliability but still within Spec | Scrapping of 25-50% of the production.	**Major** – Impact of the end-effect at the System level can be injury or impairment that requires professional medical intervention
2	Process Delay | Scrapping of 5-25% of the production. | Minor cosmetic or usability impact but still within Spec	**Minor** – Impact of the end-effect at the System level can be temporary impairment or injury that does not require professional medical intervention
1	Scrapping of 0-5% of the production. Some of the products have to be reworked	**Negligible** – Impact of the end-effect at the System level can be at most an inconvenience, or temporary discomfort

RPN	Action
53-125	**Level 3** - Reduce RPN through failure compensating provisions.
13-52	**Level 2** - If Safety Impact is Y, reduce RPN to as low as possible. If Safety Impact is N, reduce RPN if feasible.
1-12	**Level 1** - If Safety Impact is Y, reduce RPN to as low as possible. If Safety Impact is N, further RPN reduction is not required.

Probability of Occurrence Criteria (Occ)			
Category	**Rank**	**Qualitative Criteria**	**Quantitative Criteria**
Frequent	5	The occurrence is frequent. Failure may be almost certain or constant failure.	$\geq 10^{-1}$
Probable	4	The occurrence is probable. Failure may be likely | Repeated failures are expected.	$< 10^{-1}$ and $\geq 10^{-2}$
Occasional	3	The occurrence is occasional | Failures may occur at infrequent intervals.	$< 10^{-2}$ and $\geq 10^{-3}$
Remote	2	The occurrence is remote | Failures are seldom expected to occur.	$< 10^{-3}$ and $\geq 10^{-4}$
Improbable	1	The occurrence is improbable | Failure is not expected to occur.	$< 10^{-4}$

Detection Criteria (Det)			
Category	**Rank**	**Qualitative Criteria**	**Quantitative Criteria**
Undetectable	5	Physics-of-Failure not understood | No detection opportunity (e.g. no inspection, or no means for detection)	$< 10^{-3}$
Low	4	Failure is very difficult to detect | Opportunity for detection is low (e.g., very low sampling for inspection)	$< 10^{-2}$ and $\geq 10^{-3}$
Moderate	3	Failure is moderately detectable | Opportunity for detection is moderate (e.g. 10% sampling)	$< 10^{-1}$ and $\geq 10^{-2}$
High	2	Failure is very detectable | Opportunity for detection is high (e.g. 100% visual inspection)	$< 9 \times 10^{-1}$ and $\geq 10^{-1}$
Almost Certain	1	Failure is obvious | Opportunity for detection is almost certain (e.g., 100% instrumented inspection)	$\geq 9 \times 10^{-1}$

Example PFMEA - Copyright Bijan Elahi 2018

BXM

PFMEA
Vivio AED

Doc # 12345
Revision 1.0

Revision History

Revision	Author	CR	Description of Change
1.0	John Adams	N/A	First approved version

Example PFMEA - Copyright 2018 Bijan Elahi

BXM

PFMEA - Vivio AED
Log of Working Sessions

Doc # 12345
Revision 1.0

Date	Participants
2017/1/10	John Adams, James Polk, Abe Lincoln, John Kennedy
2017/1/25	John Adams, James Polk, Abe Lincoln, John Kennedy
2017/2/16	John Adams, James Polk, Abe Lincoln, John Kennedy, Michael Jackson

Example PFMEA - Copyright 2018 Bijan Elahi

C.10 USE/MISUSE FAILURE MODES AND EFFECTS ANALYSIS

BXM

UMFMEA
Vivio AED

Doc # 12346
Revision 1.0

Introduction

Use-Misuse Failure Modes and Effects Analysis (UMFMEA) analyses failures that are related to use by the User. UMFMEA also considers potential misuses. Abnormal use or malice are excluded.

Reasonably Foreseeable Misuses are also analyzed in this analysis. Misuse is not use failure. It is deliberate and well-intentioned. Example: Off-label use

System Under Analysis:

Vivio AED model 1234, version 1.1

Primary functions:

Detect VF and deliver therapeutic shock

Secondary functions:

Monitor AED health and report any abnormalities
Monitor the connections of the defibrillation pads, both to Vivio and to patient's

Scope

The scope of this analysis is the interactions between the user and Vivio AED. The figure below depicts the use scenarios that are applicable to this analysis.

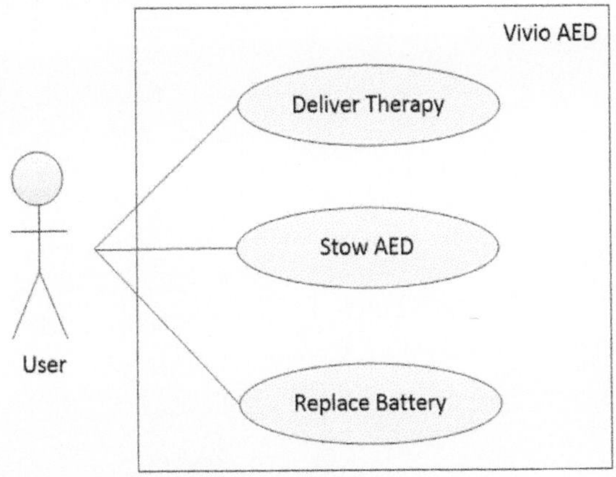

BXM

UMFMEA
Vivio AED

Doc # 12346
Revision 1.0

Use Scenario		POTENTIAL FAILURE MODES & EFFECTS			
ID	Step Action	Failure Mode	Causes/Mechanisms of Failure	Local Effects of Failure	End Effects of Failure
Deliver Therapy					
Task 1 - Prepare AED					
1	Remove AED from outer cover	Unable to open AED outer cover	Zipper too tight	N/A	No therapy
2	Press green On/Off button	Fail to press the green on/off button	Cognition error → device not started → defib shock not delivered	N/A	No therapy
Task 2 - Prepare Patient					
3	Expose patient's chest	Don't expose patient's chest	Unable to remove clothing → Unable to attach defib pads → no shock delivered	N/A	No therapy
4	Clean and ready skin for pads	Leave hair or dirt on patient's skin	Unable to clean and prepare skin → defib pads partially adhere to skin → Full shock energy is not delivered to patient	N/A	Inadequate therapy
5	Clean and ready skin for pads	Leave hair or dirt on patient's skin	Unable to clean and prepare skin → defib pads partially adhere to skin → sparks between pads and skin	N/A	Sparks (pad to skin)

Example UMFMEA - Copyright 2018 Bijan Elahi

BXM

UMFMEA
Vivio AED

Doc # 12346
Revision 1.0

		INITIAL RATING					FINAL RATING				
Safety Impact?	Existing Mitigations	Sev	Occ	Det	RPN (auto)	Additional Mitigations	Sev	Occ	Det	RPN (auto)	Remarks
Y	N/A	5	3	1	15	Use of plastic, self lubricating zipper	5	2	1	10	
Y	Use of audio commands to user	5	2	2	20	* Use of flashing green light to alert user	5	1	2	10	
Y	N/A	5	3	1	15	Scissors are supplied in the Vivio outer case	5	1	1	5	
Y	audio visual guidance on UI	4	3	1	12	Razor and alcohol wipes supplied in the Vivio outer case	5	2	1	10	
Y	audio visual guidance on UI	3	3	1	9	* SW monitors EKG signal quality and warns of bad contact * Razor and alcohol wipes supplied in the Vivio outer case	3	2	1	6	

BXM

UMFMEA
Vivio AED

Doc # 12346
Revision 1.0

Use Scenario		POTENTIAL FAILURE MODES & EFFECTS			
ID	**Step Action**	**Failure Mode**	**Causes/Mechanisms of Failure**	**Local Effects of Failure**	**End Effects of Failure**
Task 3 - Apply Defib. Pads					
6	Remove pads from pouch	Don't remove pads from pouch	Cognition error	N/A	No therapy
7	Adhere pads to patient's chest according to the guidance on AED display, or the pads pouch	Improperly place the pads on patient's chest	Perception error - cannot see visual information	N/A	Inadequate therapy
8	Adhere pads to patient's chest according to the guidance on AED display, or the pads pouch	Contaminate pad adhesive prior to application to patient's skin	Action error - contamination of adhesive → reduced adhesion of pads to skin	N/A	Inadequate therapy
Task 4 - Deliver Shock					
9	Follow audio guidance: do not touch patient while AED analyzes the heart rhythm	Touch/move patient while EKG analysis is ongoing	Perception error - user doesn't hear/see instructions to not touch the patient → False positive VF detection	N/A	Inappropriate shock
10	Press orange shock button if audio guidance instructs so	Do not press orange button to initiate shock	Perception error - user doesn't hear/see instructions to press the orange shock button → no shock delivered	N/A	No therapy

Example UMFMEA - Copyright 2018 Bijan Elahi

BXM

UMFMEA
Vivio AED

Doc # 12346
Revision 1.0

		INITIAL RATING					FINAL RATING				
Safety Impact?	Existing Mitigations	Sev	Occ	Det	RPN (auto)	Additional Mitigations	Sev	Occ	Det	RPN (auto)	Remarks
Y	Use of audio and visual commands to user	5	2	1	10	N/A	5	2	1	10	
Y	N/A	4	3	2	24	* Large-print diagram on the pad pouch * Animated graphics on LCD display * Audio guidance	4	2	1	8	
Y	N/A	4	3	3	36	Hands-free applicator of defib pads	4	1	3	12	
Y	Use of both audio and visual communication	3	2	1	6	N/A	3	2	1	6	
Y	Use of both audio and visual communication	5	2	1	10	N/A	5	2	1	10	

BXM

UMFMEA
Vivio AED

Doc # 12346
Revision 1.0

Use Scenario		POTENTIAL FAILURE MODES & EFFECTS			
ID	**Step Action**	**Failure Mode**	**Causes/Mechanisms of Failure**	**Local Effects of Failure**	**End Effects of Failure**
Stow AED					
Task 1 - Check AED					
11	Check for damage to AED	Do not check for damage to AED	User forgets to follow protocol	N/A	Potentially damaged device may go unnoticed
12	Check for dirt or contamination of AED	Do not check for contamination of AED	User forgets to follow protocol	N/A	Potentially damaged device may go unnoticed
13	Clean AED, if necessary per guidelines in IFU	Clean AED with unapproved chemicals	User uses harsh chemical to clean AED → damage to AED surface finish → LCD screen becomes dull	N/A	Future use of the device may be hampered
14	Clean AED, if necessary per guidelines in IFU	Submerges AED in water for cleaning	User submerges AED in water for cleaning → AED is damaged and becomes non-functional	N/A	AED will become non-functional
15	Plug the cable connector for a new pad-set into Vivio (do not open the pads case)	Do not plug cable connector for the new pad-set	Lapse → user forgets to plug in the new pads' connector into the AED	N/A	At the next use, the user will have to plug in the connector before using the AED
16	Check supplies and accessories for damage and expiration dates	Do not check expired or damaged supplies	User forgets to follow protocol	N/A	Potentially damaged or expired supplies may go unnoticed
17	Replace any damaged or expired supply	Do not replace damaged or expired supplies	User forgets to follow protocol	N/A	Potentially damaged or expired supplies may go unnoticed

Example UMFMEA - Copyright 2018 Bijan Elahi

BXM

UMFMEA
Vivio AED

Doc # 12346
Revision 1.0

		INITIAL RATING					FINAL RATING				
Safety Impact?	Existing Mitigations	Sev	Occ	Det	RPN (auto)	Additional Mitigations	Sev	Occ	Det	RPN (auto)	Remarks
N	Instructions in IFU	3	3	1	9	N/A	3	3	1	9	
N	Instructions in IFU	3	3	1	9	N/A	3	3	1	9	
N	Instructions in IFU	3	3	1	9	N/A	3	3	1	9	
N	Instructions in IFU	4	2	1	8	N/A	4	2	1	8	
N	N/A	4	3	1	12	AED senses that the pads have not been plugged in and will continuously chirp until they are	4	2	1	8	
N	Instructions in IFU	3	3	1	9	N/A	3	3	1	9	
N	Instructions in IFU	3	3	1	9	N/A	3	3	1	9	

Example UMFMEA - Copyright 2018 Bijan Elahi

BXM

UMFMEA
Vivio AED

Doc # 12346
Revision 1.0

Use Scenario		POTENTIAL FAILURE MODES & EFFECTS			
ID	**Step Action**	**Failure Mode**	**Causes/Mechanisms of Failure**	**Local Effects of Failure**	**End Effects of Failure**
Task 2 - Check Status					
18	Remove battery for 5+ seconds	Do not remove battery	User forgets to follow protocol	N/A	Low battery may go unnoticed
19	Remove battery for 5+ seconds	Remove and replace battery in less than 5 Sec.	User forgets to follow protocol	N/A	Low battery may go unnoticed
20	Reinsert battery and go through self-test procedure	Reinsert battery but not go through self-tests	User forgets to follow protocol	N/A	Problems with the AED may go unnoticed
Task 3 - Return AED to its Storage Location					
21	Close the outer case and place AED in its storage location	Do not close the outer case	User forgets to follow protocol	N/A	AED is exposed to elements
22	Close the outer case and place AED in its storage location	Do not return AED to its storage place	User forgets to follow protocol	N/A	AED may not be available for next use

Example UMFMEA - Copyright 2018 Bijan Elahi

BXM

UMFMEA
Vivio AED

Doc # 12346
Revision 1.0

		INITIAL RATING					FINAL RATING				
Safety Impact?	Existing Mitigations	Sev	Occ	Det	RPN (auto)	Additional Mitigations	Sev	Occ	Det	RPN (auto)	Remarks
N	N/A	4	3	1	12	AED performs self-tests everyday and alert user incase of low battery	4	2	1	8	
N	N/A	4	3	1	12	AED performs self-tests everyday and alert user incase of low battery	4	2	1	8	
N	N/A	4	3	1	12	AED performs self-tests everyday and alert user incase of any failures	4	2	1	8	
N	N/A	4	3	1	12	AED performs self-tests everyday and alert user incase of any failures	4	2	1	8	
N	N/A	1	2	1	2	N/A	1	2	1	2	

Example UMFMEA - Copyright 2018 Bijan Elahi

BXM

UMFMEA
Vivio AED

Doc # 12346
Revision 1.0

Use Scenario		POTENTIAL FAILURE MODES & EFFECTS			
ID	**Step Action**	**Failure Mode**	**Causes/Mechanisms of Failure**	**Local Effects of Failure**	**End Effects of Failure**
Replace Battery					
Task 1 - Obtain and replace battery					
23	Use recommended replacement battery	Obtain unapproved knock-off battery	Cost savings or failure to follow IFU → Secondary market battery is purchased	N/A	Battery may not last as long
24	Take out old battery	Do not take out old battery	User forgets to take out old battery → User thinks battery has been replaced	N/A	Battery may be depleted soon
25	Insert new battery	Do not insert new battery	User removes old battery but forgets to insert new battery	N/A	AED will become non-functional
26	Go through self-test checks	Do not go through self-tests	User forgets to follow protocol	N/A	Problems with the AED may go unnoticed
27	Dispose old battery	Throw old battery in trash	User forgets to follow protocol	N/A	Violation of local regulations
Misuses					
28	Use Vivio AED per labeled and approved indications	Use Vivio AED on a child or infant	Emergency situation for a child → User decides to use the AED off-label	N/A	Potential permanent physical injury to child

Example UMFMEA - Copyright 2018 Bijan Elahi

BXM

UMFMEA
Vivio AED

Doc # 12346
Revision 1.0

		INITIAL RATING					FINAL RATING				
Safety Impact?	**Existing Mitigations**	**Sev**	**Occ**	**Det**	**RPN (auto)**	**Additional Mitigations**	**Sev**	**Occ**	**Det**	**RPN (auto)**	**Remarks**
N	Instructions in IFU	1	3	2	6	N/A	1	3	2	6	
N	N/A	3	2	3	18	AED performs self-tests everyday and alert user incase of low battery	3	2	1	6	
N	Instructions in IFU	5	1	1	5	N/A	5	1	1	5	
N	N/A	4	3	1	12	AED performs self-tests everyday and alert user incase of any failures	4	2	1	8	
Y	Instructions in IFU	1	2	1	2	N/A	1	2	1	2	
Y	Clear cautions and warning on the device and in the IFU that Vivio AED is not intended for children	4	2	1	8	N/A	4	2	1	8	

Severity Criteria (Sev)

Rank	Severity Descriptions - No Safety Impact	Severity Description -- Safety Impact
5	**Catastrophic**: Described failure mode will cause immediate failure of the Subject. (Total loss of all function – primary and secondary)	**Catastrophic:** Impact of the end-effect at the System level can be death
4	**Critical**: Described failure mode will severely impact Subject functionality | Complete loss of primary functions	**Critical:** Impact of the end-effect at the System level can be permanent impairment or life-threatening injury
3	**Serious:** Described failure mode will reduce Subject functionality. (Partial loss of primary function | Complete loss of secondary functions)	**Serious:** Impact of the end-effect at the System level can be injury or impairment that requires professional medical intervention
2	**Minor**: Described failure mode will have temporal or self-restoring impact on functionality | partial loss of secondary functions	**Minor:** Impact of the end-effect at the System level can be temporary injury or impairment that does not require professional medical intervention
1	**None**: Described failure mode will have no impact on functionality | Annoyance / inconvenience of the user	**Negligible:** Impact of the end-effect at the System level can be at most an inconvenience, or temporary discomfort

Probability of Occurrence Criteria (Occ)

Category	Rank	Qualitative Criteria
Frequent	5	The occurrence is frequent. Experienced by almost every user.
Probable	4	The occurrence is probable. Experienced by most users.
Occasional	3	The occurrence is occasional. Experienced by some users.
Remote	2	The occurrence is remote. Experienced by few users.
Improbable	1	The occurrence is improbable. Has not been observed; not expected to be experienced by any user.

RPN	Action
53-125	**Level 3** - Reduce RPN through failure compensating provisions.
13-52	**Level 2** - If Safety Impact is Y, reduce RPN to as low as possible. If Safety Impact is N, reduce RPN if feasible.
1-12	**Zone 1** - If Safety Impact is Y, reduce RPN to as low as possible. If Safety Impact is N, further risk reduction is not required.

Detection Criteria (Det)

Category	Rank	Descriptions
Undetectable	5	Effect is not immediately visible or knowable | Countermeasures not possible
Low	4	Effect can be visible or knowable only with expert investigation using specialized equipment | Countermeasures are unlikely
Moderate	3	Effect can be visible or knowable with the moderate effort by user | Countermeasures are probable
High	2	Highly Detectable - Effect can be visible or knowable with simple action by user, from the information provided by the system itself | Countermeasures are likely
Almost Certain	1	Almost certain detection - Effect is clearly visible or knowable to user without any further action by user | Countermeasures are certain

Example UMFMEA - Copyright 2018 Bijan Elahi

BXM

UMFMEA
Vivio AED

Doc # 12346
Revision 1.0

Revision History

Revision	Author	CR	Description of Change
1.0	John Adams	N/A	First approved version

Example UMFMEA - Copyright 2018 Bijan Elahi

BXM

UMFMEA - Vivio AED
Log of Working Sessions

Doc # 12346
Revision 1.0

Date	Participants
2017/2/11	John Adams, David Souter, Sam Alito, John Kennedy
2017/2/25	John Adams, David Souter, Sam Alito, John Kennedy
2017/3/17	John Adams, David Souter, Sam Alito, John Kennedy, Michael Jackson

Example UMFMEA - Copyright 2018 Bijan Elahi

C.11 RISK ASSESSMENT AND CONTROLS TABLE

BXM

RACT - Per Hazard
Vivio EAD

Doc # 12347
Revision 1.0

ID	Hazard Source	Initial cause of hazard	Sequence of Events	Hazard	Hazardous Situations	SD
1	FTA	SW doesn't sense VF	SW doesn't sense VF → VF not detected → Therapy not delivered	Haz.01 No Therapy	Fibrillating patient does not receive therapy	VF detection algorithm xyz
2	DFMEA ID11	Pads wire connector plug slips out of socket	Pad connector plug slips out of socket → AED not connected to defib pads → shock not delivered	Haz.01 No Therapy	Fibrillating patient does not receive therapy	SW detects disconnected pad connector
3	DFMEA ID1	Battery defective	Battery defective → Power supply failed → Energy unavailable → Therapy not delivered	Haz.01 No Therapy	Fibrillating patient does not receive therapy	Vivio SW does battery health check upon installation
4	DFMEA ID8	Gate line opens	Gate line opens → FET switch not activated → Shock not delivered	Haz.01 No Therapy	Fibrillating patient does not receive therapy	Redundant gate connection with welded wires.
5	DFMEA ID10	Gate line opens	Gate line opens → FET switch not activated → Sensing circuits are not connected to defib pads → VF not detected	Haz.01 No Therapy	Fibrillating patient does not receive therapy	Redundant gate connection with welded wires.
6	DFMEA ID3	Circuit failure	Excess current → MOSFET failure → Inability to provide high voltage to shock capacitor → Therapy not delivered	Haz.01 No Therapy	Fibrillating patient does not receive therapy	Current limiter prevents over-current into MOSFET
7	DFMEA ID15	UI Button aging	UI button aging → button stuck in open/closed mode → shock not delivered	Haz.01 No Therapy	Fibrillating patient does not receive therapy	Use of high reliability switches
12	DFMEA ID17	Too much power to loudspeaker	Amplifier gain too high → Too much power to loudspeaker → blown voice coil → distorted output → User misunderstands audio instructions → improper application of defib pads	Haz.01 No Therapy	Fibrillating patient does not receive therapy	Audio drivers designed to prevent overpowering the loudspeaker
15	PFMEA ID3	ESD shock to PCB while installing	Worker/workstation are not properly grounded → Worker handles PCB and caused an ESD discharge and damage to PCB → FET switch failure → Shock capacitor is not charged	Haz.01 No Therapy	Fibrillating patient does not receive therapy	N/A
16	DFMEA ID7	Corrosion of leads	Corrosion of leads → leakage of current → loss of capacitor charge	Haz.01 No Therapy	Fibrillating patient receives inadequate shock energy for defibrillation	Capacitor leads are hermetically sealed to prevent corrosion

Example RACT - Copyright 2018 Bijan Elahi

BXM

RACT - Per Hazard
Vivio EAD

Doc # 12347
Revision 1.0

Risk Controls				P2					Risk				
PM	IS	P1	Harm	Cat	Crit	Ser	Minr	Negl	Cat	Crit	Ser	Minr	Negl
High quality conductive pad adhesive	audio visual guidance on UI for proper use of AED	1E-05	Harm.4 Persistent fibrillation	0.85	0.1	0.05	0	0	8.5E-06	1.0E-06	5.0E-07	0.0E+00	0.0E+00
Positive click for haptic feedback upon connection	audio visual guidance on UI	1E-05	Harm.4 Persistent fibrillation	0.85	0.1	0.05	0	0	8.5E-06	1.0E-06	5.0E-07	0.0E+00	0.0E+00
Periodic battery checks & alarm if battery has failed	recomm. to buy quality batteries	1E-05	Harm.4 Persistent fibrillation	0.85	0.1	0.05	0	0	8.5E-06	1.0E-06	5.0E-07	0.0E+00	0.0E+00
N/A	N/A	1.E-06	Harm.4 Persistent fibrillation	0.85	0.1	0.05	0	0	8.5E-07	1.0E-07	5.0E-08	0.0E+00	0.0E+00
N/A	N/A	1.E-06	Harm.4 Persistent fibrillation	0.85	0.1	0.05	0	0	8.5E-07	1.0E-07	5.0E-08	0.0E+00	0.0E+00
N/A	N/A	1.E-06	Harm.4 Persistent fibrillation	0.85	0.1	0.05	0	0	8.5E-07	1.0E-07	5.0E-08	0.0E+00	0.0E+00
Periodic self check; AED warns of failure PRIOIR to use of the deivce	N/A	1E-05	Harm.4 Persistent fibrillation	0.85	0.1	0.05	0	0	8.5E-06	1.0E-06	5.0E-07	0.0E+00	0.0E+00
N/A	Audio/visual and IFU instructions for proper application of defib pads	1.E-05	Harm.4 Persistent fibrillation	0.85	0.1	0.05	0	0	8.5E-06	1.0E-06	5.0E-07	0.0E+00	0.0E+00
* Require ESD wrist strap * Grounded workstation * Ion fans on work area	Training and instruction to factory workers	1.E-06	Harm.4 Persistent fibrillation	0.85	0.1	0.05	0	0	8.5E-07	1.0E-07	5.0E-08	0.0E+00	0.0E+00
N/A	N/A	1.E-06	Harm.4 Persistent fibrillation	0.85	0.1	0.05	0	0	8.5E-07	1.0E-07	5.0E-08	0.0E+00	0.0E+00

Example RACT - Copyright 2018 Bijan Elahi

BXM

RACT - Per Hazard
Vivio EAD

Doc # 12347
Revision 1.0

ID	Hazard Source	Initial cause of hazard	Sequence of Events	Hazard	Hazardous Situations	SD
17	UMFMEA ID4	User unable to clean and prepare skin for defib pad application	Unable to clean and prepare skin → defib pads partially adhere to skin → Full shock energy is not delivered to patient	Haz.01 No Therapy	Fibrillating patient receives inadequate shock energy for defibrillation	N/A
18	UMFMEA ID7	User doesn not see visual information	Perception error - cannot see visual information → Improperly place the pads on patient's chest	Haz.01 No Therapy	Fibrillating patient receives inadequate shock energy for defibrillation	N/A

→

Example RACT - Copyright 2018 Bijan Elahi

BXM

RACT - Per Hazard
Vivio EAD

Doc # 12347
Revision 1.0

Risk Controls				P2					Risk				
PM	IS	P1	Harm	Cat	Crit	Ser	Minr	Negl	Cat	Crit	Ser	Minr	Negl
Razor and alcohol wipes supplied in the Vivio outer case	Audio/visual and IFU instruct to prepare skin	1.E-05	Harm.4 Persistent fibrillation	0.85	0.1	0.05	0	0	8.5E-06	1.0E-06	5.0E-07	0.0E+00	0.0E+00
N/A	* Large-print diagram on the pad pouch * Animated graphics on LCD display * Audio guidance	1.E-05	Harm.4 Persistent fibrillation	0.85	0.1	0.05	0	0	8.5E-06	1.0E-06	5.0E-07	0.0E+00	0.0E+00
								Haz.01 No Therapy →	6.37E-05	7.50E-06	3.75E-06	0.00E+00	0.00E+00

BXM

RACT - Per Hazard
Vivio EAD

Doc # 12347
Revision 1.0

ID	Hazard Source	Initial cause of hazard	Sequence of Events	Hazard	Hazardous Situations	SD
11	FTA	Insufficient SW specificity	Insufficient SW specificity → False positive VF detection → Inappropriate shock	Haz.02 Inappropriate Shock	Patient without VF receives defibrillation shock	VF detection algorithm xyz
13	DFMEA ID17	Too much power to loudspeaker	Amplifier gain too high → Too much power to loudspeaker → blown voice coil → distorted output → User misunderstands audio instructions → improper application of defib pads	Haz.02 Inappropriate Shock	Patient without VF receives defibrillation shock	Audio drivers designed to prevent overpowering the loudspeaker
14	DFMEA ID12	Weak contact between plug and socket on defib pad connector	Weak contact between plug and socket → intermittent electrical contact on defib pad connector → noise on VF sensing → False positive VF detection	Haz.02 Inappropriate Shock	Patient without VF receives defibrillation shock	1 N spring force to retain plug
19	UMFMEA ID9	User doesn not see visual information	Perception error - user doesn't hear/see instructions to not touch the patient → Touch/move patient while EKG analysis is ongoing → False positive VF detection	Haz.02 Inappropriate Shock	Patient without VF receives defibrillation shock	N/A
				→		
8	DFMEA ID5	Circuit failure	Loss of feedback loop to µC → Charging circuit charges indefinitely → Increased internal temperature	Haz.03 Hot surfaces	User or bystander comes in contact with hot device	Thermistor feedback interlock to prevent runaway charging
9	UMFMEA ID5	Leave hair or dirt on patient's skin	Unable to clean and prepare skin → defib pads partially adhere to skin → sparks between pads and skin	Haz.04 Sparks (pad to skin)	Patient exposed to high voltage spark to skin	SW monitors EKG signal quality and warns of bad contact
10	DFMEA ID18	Impact to casing	Impact to casing → cracked AED casing → sharps on AED body	Haz.06 Sharps	User or bystander comes in contact with a sharp edge/point	Casing made of ABS to withstand high impact
20	FTA	Insulation breach	Pad wires' insulation is breached → exposure of high-voltage wires	Haz.09 Current Leakage	User is exposed to high voltage during defibrillation	Use of high-integrity components

Example RACT - Copyright 2018 Bijan Elahi

BXM

RACT - Per Hazard
Vivio EAD

Doc # 12347
Revision 1.0

Risk Controls				P2					Risk				
PM	IS	P1	Harm	Cat	Crit	Ser	Minr	Negl	Cat	Crit	Ser	Minr	Negl
N/A	N/A	1.E-03	Harm.9 Pain from electric shock	0	0	0.1	0.9	0	0.0E+00	0.0E+00	1.0E-04	9.0E-04	0.0E+00
N/A	Audio/visual and IFU instructions for proper application of defib pads	1.E-05	Harm.9 Pain from electric shock	0	0	0.1	0.9	0	0.0E+00	0.0E+00	1.0E-06	9.0E-06	0.0E+00
N/A	N/A	1.E-05	Harm.9 Pain from electric shock	0	0	0.1	0.9	0	0.0E+00	0.0E+00	1.0E-06	9.0E-06	0.0E+00
N/A	Use of both audio and visual communication	1.E-04	Harm.9 Pain from electric shock	0	0	0.1	0.9	0	0.0E+00	0.0E+00	1.0E-05	9.0E-05	0.0E+00
								Haz.02 Inappropriate Shock →	0.00E+00	0.00E+00	1.12E-04	1.01E-03	0.00E+00
Plastic casing as a poor heat conductor	N/A	1.E-06	Harm.1 Burns (thermal)	0	0.01	0.7	0.2	0.09	0.0E+00	1.0E-08	7.0E-07	2.0E-07	9.0E-08
Razor and alcohol wipes supplied in the Vivio outer case	audio visual guidance on UI	1.E-06	Harm.1 Burns (thermal)	0	0.01	0.7	0.2	0.09	0.0E+00	1.0E-08	7.0E-07	2.0E-07	9.0E-08
Soft grip surface to prevent dropping	N/A	1.E-05	Harm.6 Laceration	0	0.02	0.9	0.07	0.01	0.0E+00	2.0E-07	9.0E-06	7.0E-07	1.0E-07
Use of protective packaging for pads	Use of audio & visual communication to inform user to stand clear during defib shock	1.E-06	Harm.11 Electric Shock	0.008	0.04	0.111	0.041	0.8	8.0E-09	4.0E-08	1.1E-07	4.1E-08	8.0E-07

Example RACT - Copyright 2018 Bijan Elahi

BXM

RACT - Per Hazarous Situation
Vivio AED

Doc # 12347
Revision 1.0

ID	Hazard Source	Initial cause of hazard	Sequence of Events	Hazard	Hazardous Situations	Risk Controls SD	Risk Controls PM
1	FTA	SW doesn't sense VF	SW doesn't sense VF → VF not detected → Therapy not delivered	Haz.01 No Therapy	Fibrillating patient does not receive adequate therapy	VF detection algorithm xyz	High quality conductive pad adhesive
2	DFMEA ID11	Pads wire connector plug slips out of socket	Pad connector plug slips out of socket → AED not connected to defib pads → shock not delivered	Haz.01 No Therapy	Fibrillating patient does not receive adequate therapy	SW detects disconnected pad connector	Positive click for haptic feedback upon connection
3	DFMEA ID1	Battery defective	Battery defective → Power supply failed → Energy unavailable → Therapy not delivered	Haz.01 No Therapy	Fibrillating patient does not receive adequate therapy	Vivio SW does battery health check upon installation	Periodic battery checks & alarm if battery has failed
4	DFMEA ID8	Gate line opens	Gate line opens → FET switch not activated → Shock not delivered	Haz.01 No Therapy	Fibrillating patient does not receive adequate therapy	Redundant gate connection with welded wires.	N/A
5	DFMEA ID10	Gate line opens	Gate line opens → FET switch not activated → Sensing circuits are not connected to defib pads → VF not detected	Haz.01 No Therapy	Fibrillating patient does not receive adequate therapy	Redundant gate connection with welded wires.	N/A
6	DFMEA ID3	Circuit failure	Excess current → MOSFET failure → Inability to provide high voltage to shock capacitor → Therapy not delivered	Haz.01 No Therapy	Fibrillating patient does not receive adequate therapy	Current limiter prevents over-current into MOSFET	N/A
7	DFMEA ID15	UI Button aging	UI button aging → button stuck in open/closed mode → shock not delivered	Haz.01 No Therapy	Fibrillating patient does not receive adequate therapy	Use of high reliability switches	Periodic self check; AED warns of failure PRIOIR to use of the deivce
12	DFMEA ID17	Too much power to loudspeaker	Amplifier gain too high → Too much power to loudspeaker → blown voice coil → distorted output → User misunderstands audio instructions → improper application of defib pads	Haz.01 No Therapy	Fibrillating patient does not receive adequate therapy	Audio drivers designed to prevent overpowering the loudspeaker	N/A
15	PFMEA ID3	ESD shock to PCB while installing	Worker/workstation are not properly grounded → Worker handles PCB and caused an ESD discharge and damage to PCB → FET switch failure → Shock capacitor is not charged	Haz.01 No Therapy	Fibrillating patient does not receive adequate therapy	N/A	* Require ESD wrist strap * Grounded workstation * Ion fans on work area

Example RACT - Copyright 2018 Bijan Elahi

BXM

RACT - Per Hazarous Situation
Vivio AED

Doc # 12347
Revision 1.0

			P2					Risk				
IS	P1	Harm	Cat	Crit	Ser	Minr	Negl	Cat	Crit	Ser	Minr	Negl
audio visual guidance on UI for proper use of AED	1E-05	Harm.4 Persistent fibrillation	0.85	0.1	0.05	0	0	8.5E-06	1.0E-06	5.0E-07	0.0E+00	0.0E+00
audio visual guidance on UI	1E-05	Harm.4 Persistent fibrillation	0.85	0.1	0.05	0	0	8.5E-06	1.0E-06	5.0E-07	0.0E+00	0.0E+00
recomm. to buy quality batteries	1E-05	Harm.4 Persistent fibrillation	0.85	0.1	0.05	0	0	8.5E-06	1.0E-06	5.0E-07	0.0E+00	0.0E+00
N/A	1.E-06	Harm.4 Persistent fibrillation	0.85	0.1	0.05	0	0	8.5E-07	1.0E-07	5.0E-08	0.0E+00	0.0E+00
N/A	1.E-06	Harm.4 Persistent fibrillation	0.85	0.1	0.05	0	0	8.5E-07	1.0E-07	5.0E-08	0.0E+00	0.0E+00
N/A	1.E-06	Harm.4 Persistent fibrillation	0.85	0.1	0.05	0	0	8.5E-07	1.0E-07	5.0E-08	0.0E+00	0.0E+00
N/A	1E-05	Harm.4 Persistent fibrillation	0.85	0.1	0.05	0	0	8.5E-06	1.0E-06	5.0E-07	0.0E+00	0.0E+00
Audio/visual and IFU instructions for proper application of defib pads	1.E-05	Harm.4 Persistent fibrillation	0.85	0.1	0.05	0	0	8.5E-06	1.0E-06	5.0E-07	0.0E+00	0.0E+00
Training and instruction to factory workers	1.E-06	Harm.4 Persistent fibrillation	0.85	0.1	0.05	0	0	8.5E-07	1.0E-07	5.0E-08	0.0E+00	0.0E+00

Example RACT - Copyright 2018 Bijan Elahi

BXM

RACT - Per Hazarous Situation
Vivio AED

Doc # 12347
Revision 1.0

ID	Hazard Source	Initial cause of hazard	Sequence of Events	Hazard	Hazardous Situations	Risk Controls SD	Risk Controls PM
16	DFMEA ID7	Corrosion of leads	Corrosion of leads → leakage of current → loss of capacitor charge	Haz.01 No Therapy	Fibrillating patient does not receive adequate therapy	Capacitor leads are hermetically sealed to prevent corrosion	N/A
17	UMFMEA ID4	User unable to clean and prepare skin for defib pad application	Unable to clean and prepare skin → defib pads partially adhere to skin → Full shock energy is not delivered to patient	Haz.01 No Therapy	Fibrillating patient does not receive adequate therapy	N/A	Razor and alcohol wipes supplied in the Vivio outer case
18	UMFMEA ID7	User doesn not see visual information	Perception error - cannot see visual information → Improperly place the pads on patient's chest	Haz.01 No Therapy	Fibrillating patient does not receive adequate therapy	N/A	N/A

→

Example RACT - Copyright 2018 Bijan Elahi

BXM

RACT - Per Hazarous Situation
Vivio AED

Doc # 12347
Revision 1.0

			P2					Risk				
IS	P1	Harm	Cat	Crit	Ser	Minr	Negl	Cat	Crit	Ser	Minr	Negl
N/A	1.E-06	Harm.4 Persistent fibrillation	0.85	0.1	0.05	0	0	8.5E-07	1.0E-07	5.0E-08	0.0E+00	0.0E+00
Audio/visual and IFU instruct to prepare skin	1.E-05	Harm.4 Persistent fibrillation	0.85	0.1	0.05	0	0	8.5E-06	1.0E-06	5.0E-07	0.0E+00	0.0E+00
* Large-print diagram on the pad pouch * Animated graphics on LCD display * Audio guidance	1.E-05	Harm.4 Persistent fibrillation	0.85	0.1	0.05	0	0	8.5E-06	1.0E-06	5.0E-07	0.0E+00	0.0E+00
Fibrillating patient does not receive adequate therapy →								6.37E-05	7.50E-06	3.75E-06	0.00E+00	0.00E+00

Example RACT - Copyright 2018 Bijan Elahi

BXM

RACT - Per Hazarous Situation
Vivio AED

Doc # 12347
Revision 1.0

ID	Hazard Source	Initial cause of hazard	Sequence of Events	Hazard	Hazardous Situations	Risk Controls SD	Risk Controls PM
11	FTA	Insufficient SW specificity	Insufficient SW specificity → False positive VF detection → Inappropriate shock	Haz.02 Inappropriate Shock	Patient without VF receives defibrillation shock	VF detection algorithm xyz	N/A
13	DFMEA ID17	Too much power to loudspeaker	Amplifier gain too high → Too much power to loudspeaker → blown voice coil → distorted output → User misunderstands audio instructions → improper application of defib pads	Haz.02 Inappropriate Shock	Patient without VF receives defibrillation shock	Audio drivers designed to prevent overpowering the loudspeaker	N/A
14	DFMEA ID12	Weak contact between plug and socket on defib pad connector	Weak contact between plug and socket → intermittent electrical contact on defib pad connector → noise on VF sensing → False positive VF detection	Haz.02 Inappropriate Shock	Patient without VF receives defibrillation shock	1 N spring force to retain plug	N/A
19	UMFMEA ID9	User doesn not see visual information	Perception error - user doesn't hear/see instructions to not touch the patient → Touch/move patient while EKG analysis is ongoing → False positive VF detection	Haz.02 Inappropriate Shock	Patient without VF receives defibrillation shock	N/A	N/A
					→		
8	DFMEA ID5	Circuit failure	Loss of feedback loop to µC → Charging circuit charges indefinitely → Increased internal temperature	Haz.03 Hot surfaces	User or bystander comes in contact with hot device	Thermistor feedback interlock to prevent runaway charging	Plastic casing as a poor heat conductor
9	UMFMEA ID5	Leave hair or dirt on patient's skin	Unable to clean and prepare skin → defib pads partially adhere to skin → sparks between pads and skin	Haz.04 Sparks (pad to skin)	Patient exposed to high voltage spark to skin	SW monitors EKG signal quality and warns of bad contact	Razor and alcohol wipes supplied in the Vivio outer case
10	DFMEA ID18	Impact to casing	Impact to casing → cracked AED casing → sharps on AED body	Haz.06 Sharps	User or bystander comes in contact with a sharp edge/point	Casing made of ABS to withstand high impact	Soft grip surface to prevent dropping
20	FTA	Insulation breach	Pad wires' insulation is breached → exposure of high-voltage wires	Haz.09 Current Leakage	User is exposed to high voltage during defibrillation	Use of high-integrity components	Use of protective packaging for pads

Example RACT - Copyright 2018 Bijan Elahi

BXM

RACT - Per Hazarous Situation
Vivio AED

Doc # 12347
Revision 1.0

			P2					Risk				
IS	**P1**	**Harm**	**Cat**	**Crit**	**Ser**	**Minr**	**Negl**	**Cat**	**Crit**	**Ser**	**Minr**	**Negl**
N/A	1.E-03	Harm.9 Pain from electric shock	0	0	0.1	0.9	0	0.0E+00	0.0E+00	1.0E-04	9.0E-04	0.0E+00
Audio/visual and IFU instructions for proper application of defib pads	1.E-05	Harm.9 Pain from electric shock	0	0	0.1	0.9	0	0.0E+00	0.0E+00	1.0E-06	9.0E-06	0.0E+00
N/A	1.E-05	Harm.9 Pain from electric shock	0	0	0.1	0.9	0	0.0E+00	0.0E+00	1.0E-06	9.0E-06	0.0E+00
Use of both audio and visual communication	1.E-04	Harm.9 Pain from electric shock	0	0	0.1	0.9	0	0.0E+00	0.0E+00	1.0E-05	9.0E-05	0.0E+00
Patient without VF receives defibrillation shock →								0.00E+00	0.00E+00	1.12E-04	1.01E-03	0.00E+00
N/A	1.E-06	Harm.1 Burns (thermal)	0	0.01	0.7	0.2	0.09	0.0E+00	1.0E-08	7.0E-07	2.0E-07	9.0E-08
audio visual guidance on UI	1.E-06	Harm.1 Burns (thermal)	0	0.01	0.7	0.2	0.09	0.0E+00	1.0E-08	7.0E-07	2.0E-07	9.0E-08
N/A	1.E-05	Harm.6 Laceration	0	0.02	0.9	0.07	0.01	0.0E+00	2.0E-07	9.0E-06	7.0E-07	1.0E-07
Use of audio & visual communication to inform user to stand clear during defib shock	1.E-06	Harm.11 Electric Shock	0.008	0.04	0.111	0.041	0.8	8.0E-09	4.0E-08	1.1E-07	4.1E-08	8.0E-07

Example RACT - Copyright 2018 Bijan Elahi

BXM

RACT - Per Harm; Overall
Vivio EAD

Doc # 12347
Revision 1.0

ID	Hazard Source	Initial cause of hazard	Sequence of Events	Hazard	Hazardous Situations	Risk Control SD	Risk Control PM
1	FTA	SW doesn't sense VF	SW doesn't sense VF → VF not detected → Therapy not delivered	Haz.01 No Therapy	Fibrillating patient does not receive adequate therapy	VF detection algorithm xyz	High quality conductive pad adhesive
2	DFMEA ID11	Pads wire connector plug slips out of socket	Pad connector plug slips out of socket → AED not connected to defib pads → shock not delivered	Haz.01 No Therapy	Fibrillating patient does not receive adequate therapy	SW detects disconnected pad connector	Positive click for haptic feedback upon connection
3	DFMEA ID1	Battery defective	Battery defective → Power supply failed → Energy unavailable → Therapy not delivered	Haz.01 No Therapy	Fibrillating patient does not receive adequate therapy	Vivio SW does battery health check upon installation	Periodic battery checks & alarm if battery has failed
4	DFMEA ID8	Gate line opens	Gate line opens → FET switch not activated → Shock not delivered	Haz.01 No Therapy	Fibrillating patient does not receive adequate therapy	Redundant gate connection with welded wires.	N/A
5	DFMEA ID10	Gate line opens	Gate line opens → FET switch not activated → Sensing circuits are not connected to defib pads → VF not detected	Haz.01 No Therapy	Fibrillating patient does not receive adequate therapy	Redundant gate connection with welded wires.	N/A
6	DFMEA ID3	Circuit failure	Excess current → MOSFET failure → Inability to provide high voltage to shock capacitor → Therapy not delivered	Haz.01 No Therapy	Fibrillating patient does not receive adequate therapy	Current limiter prevents over-current into MOSFET	N/A
7	DFMEA ID15	UI Button aging	UI button aging → button stuck in open/closed mode → shock not delivered	Haz.01 No Therapy	Fibrillating patient does not receive adequate therapy	Use of high reliability switches	Periodic self check; AED warns of failure PRIOIR to use of the deivce
12	DFMEA ID17	Too much power to loudspeaker	Amplifier gain too high → Too much power to loudspeaker → blown voice coil → distorted output → User misunderstands audio instructions → improper application of defib pads	Haz.01 No Therapy	Fibrillating patient does not receive adequate therapy	Audio drivers designed to prevent overpowering the loudspeaker	N/A

Example RACT - Copyright 2018 Bijan Elahi

BXM

RACT - Per Harm; Overall
Vivio EAD

Doc # 12347
Revision 1.0

s			P2					Risk				
IS	P1	Harm	Cat	Crit	Ser	Minr	Negl	Cat	Crit	Ser	Minr	Negl
audio visual guidance on UI for proper use of AED	0.00001	Harm.4 Persistent fibrillation	0.85	0.1	0.05	0	0	8.5E-06	1.0E-06	5.0E-07	0.0E+00	0.0E+00
audio visual guidance on UI	0.00001	Harm.4 Persistent fibrillation	0.85	0.1	0.05	0	0	8.5E-06	1.0E-06	5.0E-07	0.0E+00	0.0E+00
recomm. to buy quality batteries	0.00001	Harm.4 Persistent fibrillation	0.85	0.1	0.05	0	0	8.5E-06	1.0E-06	5.0E-07	0.0E+00	0.0E+00
N/A	1.E-06	Harm.4 Persistent fibrillation	0.85	0.1	0.05	0	0	8.5E-07	1.0E-07	5.0E-08	0.0E+00	0.0E+00
N/A	1.E-06	Harm.4 Persistent fibrillation	0.85	0.1	0.05	0	0	8.5E-07	1.0E-07	5.0E-08	0.0E+00	0.0E+00
N/A	1.E-06	Harm.4 Persistent fibrillation	0.85	0.1	0.05	0	0	8.5E-07	1.0E-07	5.0E-08	0.0E+00	0.0E+00
N/A	0.00001	Harm.4 Persistent fibrillation	0.85	0.1	0.05	0	0	8.5E-06	1.0E-06	5.0E-07	0.0E+00	0.0E+00
Audio/visual and IFU instructions for proper application of defib pads	1.E-05	Harm.4 Persistent fibrillation	0.85	0.1	0.05	0	0	8.5E-06	1.0E-06	5.0E-07	0.0E+00	0.0E+00

Example RACT - Copyright 2018 Bijan Elahi

BXM

RACT - Per Harm; Overall
Vivio EAD

Doc # 12347
Revision 1.0

ID	Hazard Source	Initial cause of hazard	Sequence of Events	Hazard	Hazardous Situations	Risk Control	
						SD	PM
15	PFMEA ID3	ESD shock to PCB while installing	Worker/workstation are not properly grounded → Worker handles PCB and caused an ESD discharge and damage to PCB → FET switch failure → Shock capacitor is not charged	Haz.01 No Therapy	Fibrillating patient does not receive adequate therapy	N/A	* Require ESD wrist strap * Grounded workstation * Ion fans on work area
16	DFMEA ID7	Corrosion of leads	Corrosion of leads → leakage of current → loss of capacitor charge	Haz.01 No Therapy	Fibrillating patient does not receive adequate therapy	Capacitor leads are hermetically sealed to prevent corrosion	N/A
17	UMFMEA ID4	User unable to clean and prepare skin for defib pad application	Unable to clean and prepare skin → defib pads partially adhere to skin → Full shock energy is not delivered to patient	Haz.01 No Therapy	Fibrillating patient does not receive adequate therapy	N/A	Razor and alcohol wipes supplied in the Vivio outer case
18	UMFMEA ID7	User doesn not see visual information	Perception error - cannot see visual information → Improperly place the pads on patient's chest	Haz.01 No Therapy	Fibrillating patient does not receive adequate therapy	N/A	N/A
19	UMFMEA ID9	User doesn not see visual information	Perception error - user doesn't hear/see instructions to not touch the patient → Touch/move patient while EKG analysis is ongoing → False positive VF detection	Haz.02 Inappropriate Shock	Patient without VF receives defibrillation shock	N/A	N/A

Example RACT - Copyright 2018 Bijan Elahi

BXM

RACT - Per Harm; Overall
Vivio EAD

Doc # 12347
Revision 1.0

s			P2					Risk				
IS	P1	Harm	Cat	Crit	Ser	Minr	Negl	Cat	Crit	Ser	Minr	Negl
Training and instruction to factory workers	1.E-06	Harm.4 Persistent fibrillation	0.85	0.1	0.05	0	0	8.5E-07	1.0E-07	5.0E-08	0.0E+00	0.0E+00
N/A	1.E-06	Harm.4 Persistent fibrillation	0.85	0.1	0.05	0	0	8.5E-07	1.0E-07	5.0E-08	0.0E+00	0.0E+00
Audio/visual and IFU instruct to prepare skin	1.E-05	Harm.4 Persistent fibrillation	0.85	0.1	0.05	0	0	8.5E-06	1.0E-06	5.0E-07	0.0E+00	0.0E+00
* Large-print diagram on the pad pouch * Animated graphics on LCD display * Audio guidance	1.E-05	Harm.4 Persistent fibrillation	0.85	0.1	0.05	0	0	8.5E-06	1.0E-06	5.0E-07	0.0E+00	0.0E+00
		→	Harm.4 Persistent fibrillation →					6.37E-05	7.50E-06	3.75E-06	0.00E+00	0.00E+00
Use of both audio and visual communication	1.E-04	Harm.9 Pain from electric shock	0	0	0.1	0.9	0	0.0E+00	0.0E+00	1.0E-05	9.0E-05	0.0E+00
		→	Paint from electric shock →					0.00E+00	0.00E+00	1.12E-04	1.01E-03	0.00E+00

Example RACT - Copyright 2018 Bijan Elahi

BXM

RACT - Per Harm; Overall
Vivio EAD

Doc # 12347
Revision 1.0

ID	Hazard Source	Initial cause of hazard	Sequence of Events	Hazard	Hazardous Situations	Risk Control SD	Risk Control PM
8	DFMEA ID5	Circuit failure	Loss of feedback loop to μC → Charging circuit charges indefinitely → Increased internal temperature	Haz.03 Hot surfaces	User or bystander comes in contact with hot device	Thermistor feedback interlock to prevent runaway charging	Plastic casing as a poor heat conductor
9	UMFMEA ID5	Leave hair or dirt on patient's skin	Unable to clean and prepare skin → defib pads partially adhere to skin → sparks between pads and skin	Haz.04 Sparks (pad to skin)	Patient exposed to high voltage spark to skin	SW monitors EKG signal quality and warns of bad contact	Razor and alcohol wipes supplied in the Vivio outer case
10	DFMEA ID18	Impact to casing	Impact to casing → cracked AED casing → sharps on AED body	Haz.06 Sharps	User or bystander comes in contact with a sharp edge/point	Casing made of ABS to withstand high impact	Soft grip surface to prevent dropping
20	FTA	Insulation breach	Pad wires' insulation is breached → exposure of high-voltage wires	Haz.09 Current Leakage	User is exposed to high voltage during defibrillation	Use of high-integrity components	Use of protective packaging for pads

<This example RACT is truncated>

Example RACT - Copyright 2018 Bijan Elahi

BXM

RACT - Per Harm; Overall
Vivio EAD

Doc # 12347
Revision 1.0

s			P2					Risk				
IS	P1	Harm	Cat	Crit	Ser	Minr	Negl	Cat	Crit	Ser	Minr	Negl
N/A	1.E-06	Harm.1 Burns (thermal)	0	0.01	0.7	0.2	0.09	0.0E+00	1.0E-08	7.0E-07	2.0E-07	9.0E-08
audio visual guidance on UI	1.E-06	Harm.1 Burns (thermal)	0	0.01	0.7	0.2	0.09	0.0E+00	1.0E-08	7.0E-07	2.0E-07	9.0E-08
N/A	1.E-05	Harm.6 Laceration	0	0.02	0.9	0.07	0.01	0.0E+00	2.0E-07	9.0E-06	7.0E-07	1.0E-07
Use of audio & visual communication to inform user to stand clear during defib shock	1.E-06	Harm.11 Electric Shock	0.008	0.04	0.111	0.041	0.8	8.0E-09	4.0E-08	1.1E-07	4.1E-08	8.0E-07
				Overall residual risk →				6.4E-05	7.8E-06	1.3E-04	1.0E-03	1.1E-06

Example RACT - Copyright 2018 Bijan Elahi

BXM

RACT - Vivio AED
Acceptable Risk Limits

Doc # 12347
Revision 1.0

The following risk limits are derived from a survey of published data on rates of occurrence of harm to patients from the use of AEDs. This is construed as the state-of-the-art for acceptable risk levels for AED use.

R-Cat	R-Crit	R-Ser	R-Minr	R-Negl
7.1E-05	9.5E-05	1.3E-04	1.5E-03	1.0E-02

Example RACT - Copyright 2018 Bijan Elahi

BXM

RACT
Vivio AED

Doc # 12347
Revision 1.0

Abbreviations

Term	Definition
Cat	Catastrophic
Crit	Critical
ID	Identification
IS	Information for Safety
Minr	Minor
Negl	Negligible
PM	Protective Measure
RACT	Risk Assessment and Control Table
SD	Safe by Design
Ser	Serious

Revision History

Revision	Author	CR	Description of Change
1.0	John Adams	N/A	First approved version

C.12 HAZARD ANALYSIS REPORT

BXM	Hazard Analysis Report – Vivio AED	Doc # 12348 Revision 1.0 Page 1 of 9

TABLE OF CONTENTS

Example HAR – Copyright 2018 Bijan Elahi

BXM	Hazard Analysis Report – Vivio AED	Doc # 12348 Revision 1.0 Page 2 of 9

1. INTRODUCTION

This Hazard Analysis report is the summation of the outcomes of the risk management process for Vivio AED. This report is intended to serve as the repository of the risk identification, analysis, estimation and evaluation for Vivio AED and to become a part of the Risk Management File [RMF].

2. SCOPE

The scope of this analysis is the Vivio AED System, excluding the Pads, which are supplied by third party. However, the interface between the AED and the Pads is within scope.

3. RISK MANAGEMENT TEAM

The Risk Management Team is identified in [RMP] section 4.

4. PRODUCT DESCRIPTION AND INTENDED USE

Vivio is a portable Automatic External Defibrillator (AED), designed for outdoor storage, and use by minimally trained individuals. Vivio is equipped with a color LCD display. Vivio provides use-instructions in graphic format on the LCD display, and speaks in 14 languages. Language selection is configurable for use in select geographies.

When contact-pads are properly applied to the patient's chest, Vivio automatically monitors patient's heart rhythm, detects fibrillation and emits a bi-phasic trans-thoracic shock of approximately 360 J to defibrillate the heart. Vivio operates on a primary battery, which is user-replaceable. Contact-pads are single-use products which are supplied by a third-party manufacturer. Vivio is intended for use on adults, and children who are over 25 kg, or older than 8 years.

Example HAR – Copyright 2018 Bijan Elahi

BXM	Hazard Analysis Report – Vivio AED	Doc # 12348 Revision 1.0 Page 3 of 9

5. CLINICAL HAZARDS

The following is a recitation of table 1 of [CHL]. Every entry in the table is examined for applicability for Vivio AED. For hazards that are not applicable, rationales are provided. Other hazards are applicable and included in the hazard analysis for Vivio AED.

ID	Hazard	Applicability and rationale
Haz.01	No therapy	Yes. Applicable
Haz.02	Inadequate therapy	Yes. Applicable
Haz.03	Inappropriate shock	Yes. Applicable
Haz.04	Hot surfaces	Yes. Applicable
Haz.05	Sparks (pad to skin)	Yes. Applicable
Haz.06	Ionizing radiation	No. Vivio AED does not produce ionizing radiation
Haz.07	Sharps	Yes. Applicable
Haz.08	Pinch points	No. Vivio does not have pinch points.
Haz.09	Current leakage	Yes. Applicable
Haz.10	Vibration	No. Vivio is not a source of vibration.

6. KEY RISKS AND MITIGATIONS

In this section, the key hazards and associated risks which are inherent to Vivio AED are identified, and risk control strategies are described.

ID	Hazard	Strategy
Haz.01	No therapy	The main hazard related to Vivio AED is not providing therapy, when the device is expected to be functional. There are two ways in which this hazard can arise. 1) Under no fault condition – this would be due to lack of sensitivity of the VF detection algorithm. The strategy here was to improve the VF detection-algorithm sensitivity while maintaining specificity. A patented algorithm xyz, was deployed in this regard. 2) Under fault condition – this could happen due to many different reasons, e.g. use error, electronic, or mechanical faults. Multiple control measures were deployed to address all the known and foreseeable fault conditions. These risk controls are detailed in the RACT.

Example HAR – Copyright 2018 Bijan Elahi

	Hazard Analysis Report – Vivio AED	Doc # 12348
BXM		Revision 1.0
		Page 4 of 9

7. RISK ASSESSMENT AND CONTROL TABLE

Hazard identification is captured in the System [DFMEA], System [PFMEA] and the [UMFMEA]. Risk estimation and evaluation is captured in the Risk Assessment and Control Table [RACT]. The [RACT] estimates and evaluates residual risks per-hazard, per-hazardous situation, and also overall for the device. Furthermore, it identifies all the risk control measures taken to reduce risks. Risk-control option-analysis is embedded in the [RACT]. Per [RM SOP], risk controls were applied in the following order of priority:

1) Safe by design
2) Protective measures
3) Information for safety

8. RISK CONTROLS AND TRACEABILITY

The risk controls that were utilized in Vivio AED are listed and detailed in Table 1 below. Verifications of implementation and effectiveness are reflected in their corresponding columns. The entries in the Implementation column are System or subsystem requirements. The entries in the Verification columns are identification of tests that were performed to provide objective evidence of implementation and effectiveness of the risk controls. In some cases, the same test provided both the evidence of implementation and effectiveness. In those cases, the same tests references were cited under both Verification of Implementation and Effectiveness columns.

Example HAR – Copyright 2018 Bijan Elahi

BXM	Hazard Analysis Report – Vivio AED	Doc # 12348 Revision 1.0 Page 5 of 9

Table 1 – Risk Controls Traceability

Risk Control	Implementation	Verif. Implementation	Verif. Effectiveness
VF detection algorithm xyz	L1.SRS.798	DVT.VA.222 DVT.VA.223 DVT.VA.186	DVT.VA.222 DVT.VA.223 DVT.VA.186
SW detects disconnected pad connector	L1.SRS.859	DVT.VA.146	DVT.VA.146
Vivio SW does battery health check upon installation	L1.SRS.023	DVT.VA.149	DVT.VA.149
Current limiter prevents over-current into MOSFET	L2.SRS.034	SVA.00766.001	L2.SDVT.011 L2.SDVT.012
Audio drivers designed to prevent overpowering the loudspeaker	L2.SRS.088	SVA.00796.001	L2.SDVT.025 L2.SDVT.026 L2.SDVT.027
Thermistor feedback interlock to prevent runaway charging	L2.SRS.189	SVA.00556.002	L2.SDVT.132 L2.SDVT.133
SW monitors EKG signal quality and warns of bad contact	L2.SRS.245	SWT.VA.173	DVT.VA.349 DVT.VA.330
Casing made of ABS to withstand high impact	L1.SRS.099	HVA.00371.001	DVT.VA.439

Example HAR – Copyright 2018 Bijan Elahi

BXM	Hazard Analysis Report – Vivio AED	Doc # 12348 Revision 1.0 Page 6 of 9

Risk Control	Implementation	Verif. Implementation	Verif. Effectiveness
Periodic battery checks & alarm if battery has failed	L1.SRS.018	HVA.00244.001	DVT.VA.587 DVT.VA.693
Positive click for haptic feedback upon connection	L1.SRS.029	DVT.VA.111	DVT.VA.111
Audio visual guidance on UI for proper use of AED	L1.SRS.029	DVT.VA.745 DVT.VA.746	UT.VA.029 UT.VA.030 UT.VA.031

9. OVERALL RESIDUAL-RISK EVALUATION

The overall residual risk for Vivio AED was evaluated and deemed to be acceptable. The basis of this evaluation was comparison with state-of-the-art risk levels as defined in [ISO 14971], and specified in the [RMP] per the guidelines in [ISO 14971] section D.4.

Table 2 reflects the computed overall residual risk per [RACT] as compared to the acceptable risk limits as specified in the [RMP].

Table 2 – Overall Residual Risk Evaluation

	Risk Class				
	Catastrophic	Critical	Serious	Minor	Negligible
Overall Residual Risk	6.4×10^{-5}	7.8×10^{-6}	1.3×10^{-4}	1.0×10^{-3}	1.1×10^{-6}
Acceptable Risk Limits	$\leq 7.1 \times 10^{-5}$	$\leq 9.5 \times 10^{-5}$	$\leq 1.3 \times 10^{-4}$	$\leq 1.5 \times 10^{-3}$	$\leq 1.0 \times 0^{-1}$

Example HAR – Copyright 2018 Bijan Elahi

BXM	Hazard Analysis Report – Vivio AED	Doc #	12348
		Revision	1.0
		Page	7 of 9

10. BENEFIT-RISK ANALYSES

The overall residual risk, as well as the individual risks for Vivio AED were evaluated to be acceptable. The basis of this evaluation was comparison with state-of-the-art risk levels as defined in [ISO 14971], and specified in the [RMP]. The residual risks in all severity classes were computed and shown to be at or below the state-of-the-art.

Vivio AED delivers comparable benefit to state-of-the-art AEDs in the market, at a lower overall risk. It is therefore concluded that the benefits of Vivio AED outweigh its risks.

11. CONCLUSION

Vivio AED was analyzed for safety-risks according to [RMP] and in compliance with [ISO 14971]. Relevant hazards were identified, and corresponding risks were evaluated. Risk control measures were identified, implemented and verified for implementation and effectiveness. The overall residual risk of Vivio AED was evaluated to be acceptable, and the benefits of Vivio AED were deemed to outweigh its risks.

12. RISK MANAGEMENT PROCESS

The risk management process that was followed for Vivio AED is described in [RM SOP].

13. TOOLS AND METHODS

For the Preliminary Hazard Analysis, Fault Tree Analysis (FTA) was used as a tool for top-down deductive analysis. After the details of design became available, the technique of Failure Modes and Effects Analysis (FMEA) was exercised upon the design, process and user interactions with the System. The outcomes of the FMEAs were rolled up into the RACT for risk estimation and evaluation. Risk controls are also captured in the RACT.

Traceability between risk controls and verifications of implementation and effectiveness is captured in section 8 above.

Example HAR – Copyright 2018 Bijan Elahi

BXM	Hazard Analysis Report – Vivio AED	Doc # 12348 Revision 1.0 Page 8 of 9

14. PRODUCTION AND POST-PRODUCTION INFORMATION

14.1 Production Phase

Information regarding the safety of Vivio AED that was learned from the Production phase of product development has been incorporated in the [PFMEA] and rolled up into the [RACT].

14.2 Post-Production Phase

This is the premarket risk management of Vivio AED. The device has not been released to the market yet. After approval and market release, the normal surveillance, complaint handling and Clinical Evaluation activities will collect post-production information and evaluate them for inclusion in the risk management work-products.

Example HAR – Copyright 2018 Bijan Elahi

BXM	Hazard Analysis Report – Vivio AED	Doc # 12348 Revision 1.0 Page 9 of 9

15. REFERENCES / APPLICABLE DOCUMENTS

Reference	Identification	Title / additional remarks
[ISO 14971]	EN ISO 14971:2012	Application of Risk Management to Medical Devices
[PFMEA]	12345	Process Failure Modes and Effects Analysis (PFMEA)
[DFMEA]	12344	Design Failure Modes and Effects Analysis (DFMEA)
[UMFMEA]	12346	Use/Misuse Failure Modes and Effect Analysis (UMFEMA)
[RACT]	12347	Risk Assessment and Control Table
[RM SOP]	xxxxxx	Risk Management Standard Operating Procedure
[RMF]	N/A	Risk Management File
[RMP]	12340	Risk Management Plan

REVISION HISTORY

Revision	Author	CR	Description of changes
1.0	John Adams	N/A	First approved version

Example HAR – Copyright 2018 Bijan Elahi

C.13 RISK MANAGEMENT REPORT

BXM	Risk Management Report – Vivio	Doc # 12349 Revision 1.0 Page 1 of 8

Table of Contents

Example RMR – Copyright 2018 Bijan Elahi

BXM	Risk Management Report – Vivio	Doc # 12349 Revision 1.0 Page 2 of 8

1 INTRODUCTION

This Risk Management Report (RMR), which fulfills the requirements of ISO 14971 section 8, is intended to provide an overview of the outcome of the Risk Management process for Vivio AED. Details of the Risk Management activities are available in the Risk Management File [RMF].

The preliminary residual risks of the Vivio AED are captured in the [PHA], and the final residual risks are captured in the [HAR].

The Risk Management Report is a living document and will be maintained for the lifecycle of Vivio AED.

2 SCOPE

The scope of this analysis is Vivio AED, a portable class III medical device designed for outdoor storage, and use by minimally trained individuals.

Vivio AED requires the use of third-party manufactured, adhesive single-use defibrillation pads. The pads are not included in this risk analysis. However, the interface to the pads is included.

Example RMR – Copyright 2018 Bijan Elahi

BXM	Risk Management Report – Vivio	Doc # 12349 Revision 1.0 Page 3 of 8

3 CONCLUSION

The Risk Management process for Vivio AED was executed per [RM SOP] and [RMP], and achieved the following outcomes:

1. The relevant hazards were identified, and corresponding risks were analyzed. See [HAR] section 5.
2. Risk control measures were identified and implemented. See [HAR] section 8.
3. Benefit-risk analyses were performed for individual risks and overall System risks. See [HAR] section 10.
4. The overall residual risk was evaluated. See [HAR] section 9.
5. The Risk Management process was executed per the [RMP] as evidenced by the [RMT] approval of this RMR.
6. Benefits of Vivio AED outweigh its risks. See section 7, below.

4 HAZARD SUMMARY

The residual risk evaluation performed showed that all the relevant hazards are controlled such that their risks are at or below acceptable levels. See [HAR] section 9 for details.

5 SAFETY STRATEGY

Vivio AED is not a novel device. Based on historical information about the risks of the previous generations of AEDs produced by "Company", and also information about comparable systems in the market, the most significant risk of such systems is failure to detect VF and deliver therapeutic shock.

The specific safety strategy which was employed for Vivio AED was to create and deploy a new patented algorithm which increases the sensitivity of VF detection while not diminishing the specificity of VF detection. In addition, the video display of Vivio AED provides, audio/visual guidance for proper preparation and attachment of defibrillation pads to the patient chest.

BXM	Risk Management Report – Vivio	Doc # 12349 Revision 1.0 Page 4 of 8

6 OVERALL RESIDUAL-RISK EVALUATION

The overall residual risk for Vivio AED was evaluated to be acceptable. The basis of this evaluation was comparison with state-of-the-art risk levels as defined in [ISO 14971], and specified in the [RMP]. Details of the risk evaluation can be found in the [HAR]. Below a summary of overall residual risk vs. acceptable risk limits is presented.

	Risk Class				
	Catastrophic	Critical	Serious	Minor	Negligible
Overall Residual Risk	6.4×10^{-5}	7.8×10^{-6}	1.3×10^{-4}	1.0×10^{-3}	1.1×10^{-6}
Acceptable Risk Limits	$\leq 7.1 \times 10^{-5}$	$\leq 9.5 \times 10^{-5}$	$\leq 1.3 \times 10^{-4}$	$\leq 1.5 \times 10^{-3}$	$\leq 1.0 \times 0^{-1}$

7 BENEFIT-RISK ANALYSIS

The residual risks for Vivio AED were evaluated both individually, and overall. In all cases, the residual risk was at or below acceptable thresholds as established in the [RMP].

Vivio AED delivers comparable benefit to state-of-the-art AEDs in the market, at a lower overall risk. It is therefore concluded that the benefits of Vivio AED outweigh its risks.

8 RISK MANAGEMENT PROCESS SUMMARY

[RM SOP] prescribes a Risk Management Process, which is compliant to [ISO 14971]. Figure 1 depicts a representation of the risk Management Process. The [RMT] ensured compliance with [RM SOP] by reviewing and approving of the [RMF] work products, including this Risk Management Report.

Example RMR – Copyright 2018 Bijan Elahi

BXM	**Risk Management Report – Vivio**	Doc #	12349
		Revision	1.0
		Page	5 of 8

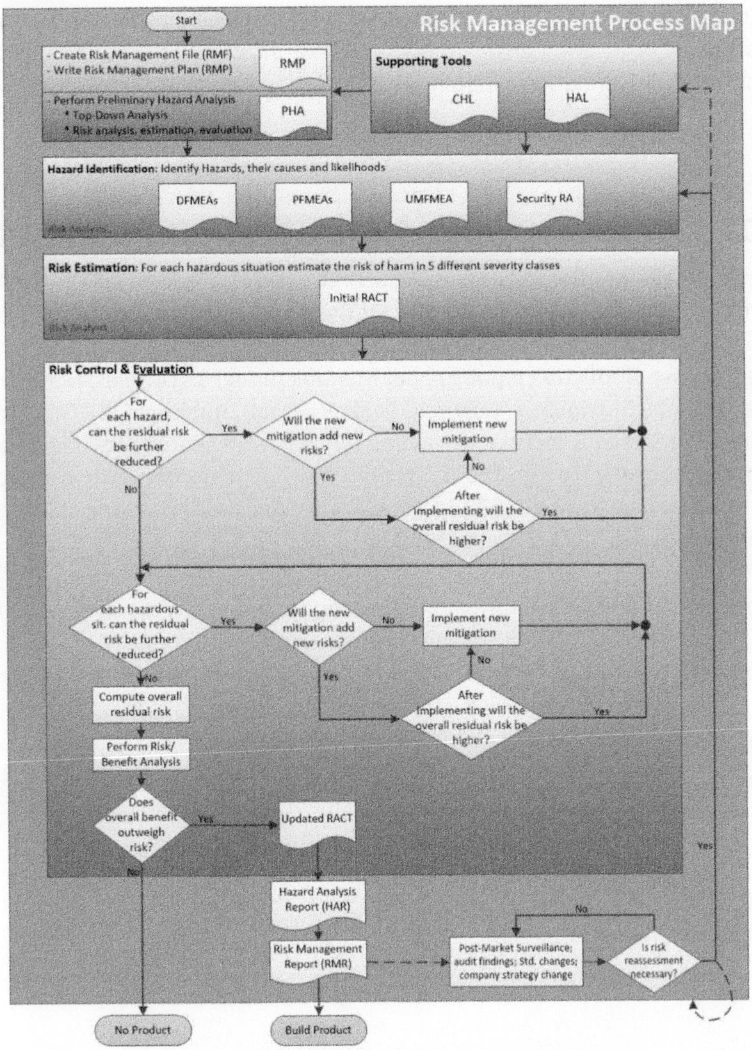

Figure 1 - Risk Management Process

BXM	Risk Management Report – Vivio	Doc # 12349 Revision 1.0 Page 6 of 8

The Risk Controls that were implemented via safety requirements were verified as part of the normal formal verification testing process. The verification results are tracked in [HAR]. Risk Controls were also validated. The validation results are tracked in [HAR] section 8.

9 PRODUCTION AND POST-PRODUCTION INFORMATION

Production feedback into the RM process is incorporated in this RMR via the PFMEAs. For Post-Production Information, per [RM SOP] the Complaint Handling, Vigilance, and Postmarket Surveillance processes are used to collect field information about Vivio AED. The sources of post-production input can be: Manufacturing, R&D, Sales, Marketing, Customers, Patients, Distributors, postmarket clinical trials, published scientific papers, news media, adverse event reports – including for competitive products. On an annual basis, or more frequently if a significant discovery is made, data from above sources is evaluated for relevance to Vivio AED. The ensuing actions depend on the collected information and can fall in a spectrum, including:

- Documentation of the information-collection actions, and discoveries where no change to the Vivio AED Risk Management artifacts is necessary
- Update to Vivio AED [RMF] including FMEAs, [HAR] and RMR with outcomes being:
    - Overall residual risk remains acceptable and benefits outweigh risks
    - Overall residual risk no longer acceptable, triggering a range of other actions e.g. Health Hazard Assessment, CAPAs, Field Corrective Actions, Vigilance reporting, redesign, etc.

Also, based on the new knowledge gained from post-production information,[CHL], [HAL], or [RM SOP] may be updated.

Example RMR – Copyright 2018 Bijan Elahi

BXM	Risk Management Report – Vivio	Doc # 12349 Revision 1.0 Page 7 of 8

10 REFERENCES

Reference	Document Number	Title / additional remarks
[RMP]	12340	Risk Management Plan
[PHA]	12341	Preliminary Hazard Analysis
[HAR]	12348	Hazard Analysis Report
[RMF]	N/A	Risk Management File. See Appendix A in this document.
[RMT]	N/A	Risk Management Team. Identified in [RMP].
[RM SOP]	xxxxxx	Risk Management Process
[ISO 14971]	EN ISO 14971:2012	Application of Risk Management to Medical Devices

11 REVISION HISTORY

Revision	Author	CR	Description of changes
1.0	John Adams	N/A	First approved version

Example RMR – Copyright 2018 Bijan Elahi

BXM	Risk Management Report – Vivio	Doc # 12349 Revision 1.0 Page 8 of 8

12 APPENDIX A – RMF INDEX

Vivio AED Risk Management File Index		
Document Title	Doc Number	Version
Risk Management Plan (RMP)	12340	1.0
Preliminary Hazard Analysis (PHA)	12341	1.0
Design Failure Modes and Effects Analysis (DFMEA)	12344	1.0
Process Failure Modes and Effects Analysis (PFMEA)	12345	1.0
Use/Misuse Failure Modes and Effects Analysis (UMFMEA)	12346	1.0
Risk Assessment and Control Table (RACT)	12347	1.0
Hazard Analysis Report (HAR)	12348	1.0
Risk Management Report (RMR)	12349	1.0
Clinical Hazards List (CHL)	12342	1.0
Harms Assessment List (HAL)	12343	1.0

Log of Post-Production Activities for Vivio AED			
Date	PMS Report No.	Version	Resulting actions and Rationale
			N/A

APPENDIX D: NBRG CONSENSUS PAPER

Notified Bodies Recommendation Group

Consensus Paper for the Interpretation and Application of Annexes Z in EN ISO 14971: 2012

Version 1.1

October 13th, 2014

Interim NBMed consensus Version

Content

Members of the workgroup:

Michael Bothe, MBA, VDE Testing & Certification Institute, Chair
Marina Belonogova, St. Jude Medical (Eucomed)
Oliver Bisazza, (COCIR)
Gert Bos, BSI
Caterina Brusasco, IBA (COCIR)
Uwe Herrmann, Roche Diagnostics (EDMA)
Hans-Heiner Junker, TÜV Süd
Peter Linders, Philips Healthcare (COCIR)
Katalin Mate, (EDMA)
Peter Meeremans, Alcon (Eucomed)
Gerd Neumann, Siemens Healthcare (COCIR)
Martin Penver, LRQA
Carrie Rydin, Terumo BCT
Thecla Sterk, (Eucomed)
Martin Schraag, Philips Healthcare, (ZVEI)
Andre Schmitz, TÜV Rheinland LGA Products GmbH
Sue Spencer, BSI
Jos van Vroonhoven, Philips Healthcare (COCIR)

1.) Introduction

In October 2010, the regular review of ISO 14971:2007 which is the basis of EN ISO 14971:2009 was closed by a broad majority of votes confirming the existing status and the wide-spread acceptance of this standard in the medical devices community, including competent authorities. In November 2010, the European Commission raised a formal objection against the use of several harmonized standards, including EN ISO 14971, followed by an in-depth assessment of the coverage of the Essential Requirements of the Medical Device Directives (90/385/EEC, 93/42/EEC and 98/79/EC) by these standards.

As a result of these objections, the Annexes Z to EN ISO 14971 were modified, resulting in EN ISO 14971:2012. This amendment of the EN ISO 14971 standard did not modify the normative parts of ISO 14971:2007[1]. The Annexes Z describe the extent of presumption of conformity that can be based on application of the normative requirements of ISO 14971 alone. The "content deviations", expressed in the revised Annexes Z, between ISO 14971:2007 and the Medical Device Directives have been commented by many experts in the field of risk management and resulted in diverging interpretations from different stakeholders (e.g. manufacturers, notified bodies, competent authorities).

This document has been prepared as a Notified Body Consensus Paper by a working group headed by the NBRG Vice Chair, with representatives from several European Notified Bodies and industry associations COCIR, Eucomed, EDMA and ZVEI. The paper aims to provide a practical interpretation of these "content deviations" to the Medical Device Directives and give guidance as to how to implement the risk management requirements. The work consolidates prior publications of various sources and is intended to facilitate common understanding between industry and Notified Bodies.

2.) Terminology

The three medical devices directives (93/42/EEC, 90/385/EEC and 98/79/EC) refer to "risk" and "safety" in a general sense. Since the time of

---

[1] EN ISO 14971:2012 Annexes Z apply to manufacturers placing devices on the market in the European Union; for the rest of the world, ISO 14971:2007 remains the applicable standard

writing these directives about 20 years ago, the knowledge of and experience in risk management have evolved considerably. This is reflected by the successive publication of EN1441 in 1994 and ISO14971-1 in 1998 as well as first (2000) and second (2007) editions of ISO 14971 which all have been recognized globally as the state-of-the-art for risk management at the moment of their publication.

The wording of the risk management aspects in the essential requirements of the Medical Devices Directives has not been modified over time, however.

The international standard ISO 14971:2007 and its European equivalent EN ISO 14971:2012 contain specific defined terms with a clear and precisely described meaning. See Table 1 for an overview of the most relevant terms used in this document. Note that these defined terms are more precise than the general terms as used in the Medical Device Directive. For example, "risk" in the medical devices directives can bear the meaning of "risk", "hazard" or "hazardous situation" depending on the context. This document is compiled on the assumption that "risk" in the Medical Device Directives is equivalent to "unacceptable risk" in ISO 14971:2007.

Term	Definition	Clause
Harm	Physical injury or damage to the health of people, or damage to property or the environment	(a) 2.2
Hazard	Potential source of harm	(a) 2.3
Hazardous situation	Circumstance in which people, property, or the environment are exposed to one or more hazard(s)	(a) 2.4
Risk	Combination of the probability of occurrence of harm and the severity of that harm	(a) 2.16
Risk control	process in which decisions are made and measures implemented by which risks are reduced to, or maintained within specified levels	(a) 2.19
Safety	Freedom from unacceptable risk	(a) 2.24
Disclosure of residual risks	Information in the accompanying documents on risks remaining after all risk control measures have been taken	(b) 5.1
Information for safety	Instructions of what actions to take or to avoid in order to prevent a hazardous situation from occurring	(b) 5.2

Table 1: Relevant terms from (a) ISO 14971:2007 and (b) ISO/TR 24971:2013

3.) General Considerations

This Consensus Paper intends to bridge the gap between the interpretation of the relevant Essential Requirements of the Medical Devices Directives, as given in the Annexes ZA, ZB, and ZC of EN ISO 14971:2012, and the practice of placing safe medical devices on the market in the EU and in other countries where the above- mentioned directives apply. This chapter provides some considerations about risk management and how this is to be interpreted in the context of the European Medical Devices Directives.

The practical approach of this Consensus Paper safeguards the principle that only medical devices that are "compatible with a high level of protection of health and safety"[2] can be placed on the EU market. With this in mind, two aspects of the European Commission's interpretation of EN ISO 14971:2012, as reflected in the Annexes Z, deserve further consideration:

1. Reduce risk "as far as possible", and
2. Economical considerations in Risk Management

1. Reducing risk "as far as possible"

The phrase "as far as possible" has led to significant confusion for those involved in placing medical devices on the market. Strict interpretation would create practical problems such as where to stop in reducing risk before a product can be placed on the market. This may restrict patient access to safe and affordable devices.

In line with Clause 1.1 of the 2013 edition of the European Commission, Parliament and Council's Joint Practical Guide of the European Parliament, the Council and the Commission for persons involved in the drafting of European Union legislation[3], this consensus paper offers Notified Bodies and manufacturers an interpretation of "as far as possible" that is "clear, easy to understand and unambiguous."

---

[2] COUNCIL DIRECTIVE 93/42/EEC of 14 June 1993 concerning medical devices.

[3] http://eur-lex.europa.eu/content/pdf/techleg/joint-practical-guide-2013-en.pdf

2. On economical considerations in Risk Management
"Content deviation" 3 in the Annexes Z of EN ISO 14971:2012 states:

a) *Annex D.8 to ISO 14971, referred to in 3.4, contains the concept of reducing risks "as low as reasonably practicable" (ALARP concept). The ALARP concept contains an element of economic consideration.*
b) *However, the* [Essential Requirements] *require risks to be reduced "as far as possible" without there being room for economic considerations.*
c) *Accordingly, manufacturers and Notified Bodies may not apply the ALARP concept with regard to economic considerations.*

This disregard of economic considerations when reducing risk is not coherent with the Medical Device Directives' objective as stated in, for example, the following recital[4] of Directive 93/42/EEC:

> Whereas the essential requirements and other requirements set out in the Annexes to this Directive, including any reference to 'minimizing' or 'reducing' risk must be interpreted and applied in such a way as to take account of technology and practice existing at the time of design and of technical and economical considerations compatible with a high level of protection of health and safety.

It should be noted that this specific recital on the relevance of economic considerations exists since the first publication of the original Directive.

4.) Recommendations for Industry

The Annexes Z of EN ISO 14971:2012 list seven "content deviations", i.e., differences between the wording of the Essential Requirements (ERs) in the Medical Device Directives and the wording of the requirements of the standard. Nevertheless, the Medical Device Directives and the standard share the same objectives of achieving a high level of product safety and ensuring continuous improvement. Parts of the content deviations are quoted and given here in *italics*. Specific recommendations for medical device manufacturers in relation to those "content deviations" are given below.

[4] Recitals are part of the legal text, as stated in Clause 10.1 of the Joint Practical Guide mentioned above and confirmed in the ECJ ruling C-219/11.

Content deviation 1: Treatment of negligible risks

"..the manufacturer must take all risks into account.."

Recommendation:

The manufacturer must identify known and foreseeable hazards and estimate the risk for each hazardous situation identified (Clause 4 of EN ISO 14971:2012). The risk control measures and the results of the risk evaluation must be recorded in the risk management file (Clause 5 and Clause 6.2 of EN ISO 14971:2012). This process ensures that all risks are given sufficient attention.

The manufacturer shall document all identified hazards and hazardous situations, their associated risks and the risk control measures for each individual risk, in the risk management file.

Compliance may be demonstrated by review of the risk management file.

Content deviation 2: Discretionary power of manufacturers as to acceptability of risks

"...all risks combined, regardless of any "acceptability assessment," need to be balanced, together with all other risks, against the benefit of the device." "Accordingly, the manufacturer may not apply any criteria of risk acceptability prior to applying Sections 1 and 2 of Annex I to Directive 93/42/EEC (Sections 1 and 6 of Annex I to Directive 90/385/EEC respectively Sections A.1 and A.2 of Annex I to Directive 98/79/EC)."

Recommendation:

When determining the criteria for risk acceptability, the manufacturer shall consider whether death or serious deterioration of health is unlikely to occur in normal operation or due to device malfunctions or deterioration of characteristics or performance, or any inadequacy in the labeling or instructions for use.[5]

---

[5] coherent with criteria from Article 10 of Directive 93/42/EEC and corresponding requirements in Annexes (e.g. Annex II, 3.1) when deciding about Field Corrective Actions

If unlikely to occur the risk shall be considered acceptable.

Otherwise, the risk must be reduced. In doing so, the manufacturer may choose an end-point for risk reduction, for example:

1.) The risk acceptability is preferably based on harmonized standards specifying state of the art risk control measures for particular categories of medical devices. Basing the risk reduction end-point on harmonized standards ensures that the risk is reduced to an acceptable level.

2.) If no harmonized standards are available, other national or international recognized standards or publications should be considered. Basing the risk reduction end-point on recognized international standards ensures that the risk is reduced to an acceptable level.

3.) Where those publications are not available, the manufacturer must assess the best risk reduction means and shall include in the description of the risk management process what criteria were used to determine the acceptability of risks. The criteria for risk acceptability are then based among others on historical data, best medical practice and state of the art.

4.) Further risk control measures do not improve the safety.

If a reduction to an acceptable level cannot be achieved, a risk-benefit analysis must demonstrate that the residual risk is outweighed by the medical benefit as explained in content deviation 4.

Compliance may be demonstrated by reflecting such end-points in the criteria for risk acceptability as part of the risk management file. Where safety cannot be demonstrated as such, existing clinical data is used to demonstrate that the medical benefit outweighs the risk.

Content deviation 3: Risk reduction "as far as possible" versus "as low as reasonably practicable"

"Essential requirements require risks to be reduced as far as possible without there being room for economic considerations..."

With this deviation the European Commission raises the concern that economic considerations might surmount safety considerations. On the other hand the reduction of a risk "as far as possible" could be without limits and the resulting devices might no longer be affordable for a larger group of patients.

Recommendation:

Although economic considerations will always be relevant in decision-making processes, the safety of the product must not be traded off against business perspectives. For transparency the manufacturer must document the end-point criteria of risk reduction based on his risk policy.

Compliance is checked by inspection of the documentation

Content deviation 4: Discretion as to whether a risk-benefit analysis needs to take place

"the manufacturer must undertake a risk-benefit analysis for the individual risk and the overall risk-benefit…"

Recommendation:

At the end of the risk management process the manufacturer shall perform a risk- benefit analysis for individual risks that are not acceptable according to the criteria explained in content deviation 2 and for which further risk reduction is not possible. In any case the manufacturer shall perform an overall risk-benefit analysis considering all individual risks to provide a rationale for overall risk acceptance.

Compliance is checked by inspection of the individual and overall risk-benefit analyses.

Content deviation 5: Discretion as to the risk control options

"…the manufacturer must apply all the control options and may not stop his endeavors if the first or second control option has reduced the risk to an "acceptable" level (unless the additional control options do not improve the safety)."

Recommendation:

As stated above for content deviation 2, the manufacturer can justify ceasing further risk reduction where it is determined that the risk is acceptable, i.e. that risk reduction has progressed to a level as described above in the section of content deviation 3.

The manufacturer shall consider all risk control measures in Essential Requirement 2 that are appropriate to reduce the risk to an acceptable level. In so doing, the manufacturer shall document the control options in the priority order, as part of the risk management process.

Compliance is checked by reviewing the documented risk management process.

Content deviation 6: Deviation as to the first risk control option

"inherent safety by design" vs. *"inherently safe design and construction"*

Recommendation:

The manufacturer shall ensure that, whenever possible, the first risk control option includes both safe design and safe construction.

Not relevant for compliance verification.

Content deviation 7: Information of the users influencing the residual risk

"..manufacturers shall not attribute additional risk reduction to the information given to the users..."

ISO 14971 (Annex J) and ISO/TR 24971 describe 'information for safety' as instructions for use, warnings, required maintenance, etc.. 'Information for safety' comprises instructions of what actions the user can take or avoid in order to prevent a hazardous situation from occurring. On the other hand, 'disclosure of residual risk' has the objective to inform users of remaining risk inherent to the use of the medical device, and concerns the risks remaining after all risk control measures have been taken.

Recommendation:

Any information for safety comprising instructions of what actions the user can take or avoid in order to prevent a hazardous situation from occurring may be considered a risk control measure. As required by Essential Requirement 13.1 of Directive 93/42/EEC (respectively ER B.8 of 98/79/EC) it may be considered as a risk control measure. The information includes the instructions for use, labels, etc.. Since 'safe use' is related to risk control measures, the Medical Device Directives do not deviate in that regard from EN ISO 14971. Any effects on risk reduction are to be documented by the manufacturer in the risk management file.

'Disclosure of residual risk' should be conducted in compliance with EN ISO 14971 Clause 6.4, 6.5 and 7. The manufacturer shall not claim a reduction to the probability of harm when disclosing residual risk.

Compliance is checked by inspection of the risk management file.

5.) Recommendations for the NB audit process

The role of the Notified Body is to assess compliance to the Directives, including the implementation of the risk management process and whether clinical benefits outweigh the risks to patients and users.

Manufacturers placing devices on the European market should be aware that gaps between the requirements of the Directives and the Risk Management Standard (documented in the EN ISO 14971: 2012 edition) have to be addressed, if applicable. It is the discretion of the manufacturer to use this standard in conjunction with other means to demonstrate conformity with the Essential Requirements of the Directive.

When EN ISO 14971 is used in upcoming audits and risk management file reviews, assessors and technical experts from Notified Bodies will focus on objective evidence on how manufacturers addressed those gaps and modified their Risk Management Process accordingly. More specifically they will evaluate:

1. Are all design solutions in conformity with the safety principles given in the Essential Requirements and EN ISO 14971 (inherent safe design and construction > protection measures > information for safety)?
2. Has the manufacturer demonstrated that all risks have been reduced to an acceptable level in the sense of this guidance paper?
3. Has the manufacturer conducted a risk benefit analysis for all individual residual risks that are not acceptable according to the risk acceptability criteria?
4. Has the manufacturer conducted an overall risk benefit analysis considering all individual risks combined?
5. Has the manufacturer demonstrated that information for safety is effective?
6. Has the manufacturer included information on residual risks, if needed, in the accompanying documents?

Annex 1: Recommendations for Standardization Bodies

Modification of IEC 60601-1 (Ed. 3.1)

Amendment A1:2012 to IEC 60601-1:2005 (together Edition 3.1) has improved the consistency of risk management terminology. Nevertheless, this edition has room for further improvement. The primary objective of the standard (and its collateral and particular standards) is to provide state-of-the-art requirements and solutions for basic safety and essential performance. The standard should be restrictive in referring to risk management.

Modification of EN ISO 14971

As has been argued, the Annexes Z to EN ISO 14971:2012 contain errors and occasionally confusing phrases. It is therefore important that these Annexes are amended through a revision of EN ISO 14971. This is within the remit of the European Standards Organizations and within CEN/CENELEC TC3 in particular. The joint Notified Bodies do not have an explicit role in this Technical Committee but, upon acceptance and implementation of this consensus paper, the NBRG is willing to contribute to amendment suggestions that subsequently may be taken forward by individual members of CEN/CENELEC TC3.

REFERENCES

[1] Guide 51, Safety Aspects—Guidelines for Their Inclusion in Standards, third ed., 2014.
[2] Regulation (EU) 2017/745 of the European Parliament and of the Council of 5 April 2017 on medical devices, amending Directive 2001/83/EC, Regulation (EC) No 178/2002 and Regulation (EC) No 1223/2009 and repealing Council Directives 90/385/EEC and 93/42/EEC.
[3] ISO 14971:2007, Medical Devices – Application of Risk Management to Medical Devices.
[4] Council Directive 93/42/EEC, Medical Device Directive (MDD).
[5] Official Journal of the European Union.
[6] Council Directive 90/385/EEC, Active Implantable Medical Device Directive (AIMDD).
[7] EN ISO 14971:2012, Medical Devices—Application of Risk Management to Medical Devices.
[8] IEC 60601-1, Edition 3.1 – Medical Electrical Equipment – Part 1: General Requirements for Basic Safety and Essential Performance.
[9] IEC 62304:2015, Medical Device Software – Software Life-Cycle Processes.
[10] P.L. Bernstein, Against the Gods: The Remarkable Story of Risk, Wiley, New York, 1998.
[11] A. Willet, The Economic Theory of Risk and Insurance, University of Pennsylvania Press, Philadelphia, PA, 1901.
[12] N.G. Leveson, Engineering a Safer World, MIT Press, Cambridge, 2012.
[13] M. Lewis, The Undoing Project, Norton, New York, 2017.
[14] IEC 62366-1:2015, Medical Devices, Part 1: Application of Usability Engineering to Medical Devices.
[15] ISO 10993-1 (2017), Biological Evaluation of Medical Devices – Part 1: Evaluation and Testing Within a Risk Management Process.
[16] MAUDE Database, FDA Manufacturer and User Facility Device Experience.
[17] EUDAMED: European Database on Medical Devices.
[18] FDA Guidance on Applying Human Factors and Usability Engineering to Medical Devices, February 3, 2016.
[19] IEC 60601-1-8:2006, Medical Electrical Equipment – Part 1-8: General Requirements for Basic Safety and Essential Performance – Collateral Standard: General Requirements, Tests and Guidance for Alarm Systems in Medical Electrical Equipment and Medical Electrical Systems.
[20] ANSI/AAMI HE 75:2009/ (R) 2013, Human Factors Engineering – Design of Medical Devices.
[21] EN ISO 13485:2016, Medical Devices—Quality Management Systems—Requirements for Regulatory Purposes.
[22] Notified Bodies Recommendation Group; Consensus Paper for the Interpretation and Application of Annexes Z in EN ISO 14971: 2012; Version 1.1; 13 Oct. 2014.
[23] M. Bordwin, Factoring the Law into Medical Device Design, MDDI, March 2005.
[24] NASA Fault Tree Handbook with Aerospace Applications; Ver 1.1, August 2002.
[25] FDA Guidance on Postmarket Management of Cybersecurity in Medical Devices, December 28, 2016.
[26] IEC TR 80002-1 Technical Report, Medical Device Software – Part 1: Guidance on the Application of ISO 14971 to Medical Device Software, Edition 1.0 2009-09.
[27] FDA Guidance for the Content of Premarket Submissions for Software Contained in Medical Devices, 2005.
[28] G.J. Holzmann, The Power of Ten – Rules for Developing Safety Critical Code, IEEE Computer, NASA/JPL Laboratory for Reliable Software, Pasadena, CA, June 2006, pp. 93–95.
[29] Factors to Consider When Making Benefit-Risk Determinations in Medical Device Premarket Approval and De Novo Classifications, FDA, March 28, 2012.
[30] Benefit-Risk Factors to Consider When Determining Substantial Equivalence in Premarket Notifications [1510(k)] with Different Technological Characteristics, FDA, July 15, 2014.

[31] Factors to Consider When Making Benefit-Risk Determinations for Medical Device Investigational Device Exemptions (IDEs), FDA, June 18, 2015.
[32] MEDDEV 2.7/1 – Clinical Evaluation: A Guide for Manufacturers and Notified Bodies Under Directives 93/42/EEC and 90/385/EEC. Revision 4, 2016.
[33] ISO 14155, Clinical Investigation of Medical Devices for Human Subjects—Good Clinical Practice, second ed., 2011.
[34] IEC TR 62366-2:2016, Medical Devices – Part 2: Guidance on the Application of Usability Engineering to Medical Devices.
[35] Code of Federal Regulations, Title 21, Part 820. <www.accessdata.fda.gov/scripts/cdrh/cfdocs/cfMAUDE/search.CFM>.
[36] R.A. Clark, R.P. Eddy, Warnings: Finding Cassandras to Stop Catastrophes, Harper Collins, New York, 2017.
[37] Christopher Chabris, Daniel Simons, The Invisible Gorilla, Broadway Paperback, New York, 2009.
[38] Simmons, Joseph & D Nelson, Leif & Simonsohn, Uri. (2011). False-Positive Psychology. Psychological science. 22. 1359–66. Available from: http://dx.doi.org/10.1177/0956797611417632.

INDEX

Note: Page numbers followed by "*f*" and "*t*" refer to figures and tables, respectively.

S

V

W

CPI Antony Rowe
Eastbourne, UK
April 19, 2021